D0765019

MICROBIOLOGY DEMYSTIFIED

Demystified Series

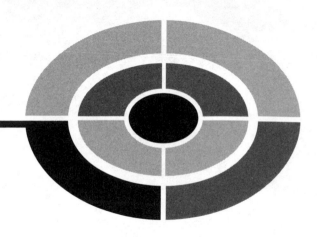

MICROBIOLOGY DEMYSTIFIED

TOM BETSY, D.C.
JIM KEOGH

McGRAW-HILL
New York   Chicago   San Francisco   Lisbon   London
Madrid   Mexico City   Milan   New Delhi   San Juan
Seoul   Singapore   Sydney   Toronto

The *McGraw·Hill* Companies

Cataloging-in-Publication Data is on file with the Library of Congress

Copyright © 2005 by The McGraw-Hill Companies, Inc. All rights reserved. Printed in the United States of America. Except as permitted under the United States Copyright Act of 1976, no part of this publication may be reproduced or distributed in any form or by any means, or stored in a data base or retrieval system, without the prior written permission of the publisher.

1 2 3 4 5 6 7 8 9 0   DOC/DOC   0 1 0 9 8 7 6 5

ISBN 0-07-144650-8

The sponsoring editor for this book was Judy Bass and the production supervisor was Pamela A. Pelton. It was set in Times Roman by Fine Composition. The art director for the cover was Margaret Webster-Shapiro; the cover designer was Handel Low.

Printed and bound by RR Donnelley.

 This book is printed on recycled, acid-free paper containing a minimum of 50% recycled, de-inked fiber.

McGraw-Hill books are available at special quantity discounts to use as premiums and sales promotions, or for use in corporate training programs. For more information, please write to the Director of Special Sales, McGraw-Hill Professional, Two Penn Plaza, New York, NY 10121-2298. Or contact your local bookstore.

I would like to dedicate this book to my wife, Shelley,
and my two babies, Juliana and Thomas,
for their encouragment during the writing of this book
and the continued joy they bring to my life.

Dr. Tom Betsy

This book is dedicated to Anne, Sandy, Joanne,
Amber-Leigh Christine, and Graaf,
without whose help and support
this book could not have been written.

Jim Keogh

CONTENTS

CONTENTS

INTRODUCTION

When you hear the words "germ," "bacteria," and "virus" you might cringe, running for the nearest sink to wash your hands. These words may bring back memories of when you caught a cold or the flu—never a pleasant experience. Germs, bacteria, viruses and other microscopic organisms are called microorganisms, or microbes for short. And as you'll learn throughout this book, some microbes cause disease while others help fight it.

Think for a moment. Right now there are thousands of tiny microbes living on the tip of your finger in a world that is so small that it can only be visited by using a microscope. In this book we'll show you how to visit this world and how to interact with these tiny creatures that call the tip of your finger home.

The microscopic world was first visited in the late 1600s by the Dutch merchant and amateur scientist Antoni van Leeuwenhoek. He was able to see living microorganisms by using a single-lens microscope. We've come a long way since Van Leeuwenhoek's first visit. Today scientists are able to see *through* some microbes and study the organelles that bring them to life.

It wasn't until the Golden Age of Microbiology between 1857 and 1914 when scientists such as Louis Pasteur and Robert Koch made a series of discoveries that rocked the scientific community. During this period scientists identified microbes that caused diseases, learned how to cure those diseases, and then prevented them from occurring through the use of immunization.

Scientists were able to achieve these remarkable discoveries by using culturing techniques to grow colonies of microbes in the laboratory. Once microbes could be grown at will, scientists focused their experiments on ways to slow that growth and stop microbes in their tracks—killing the microbe and curing the disease caused by the microbe.

Culturing microbes is central to the study of microbiology. You'll be using many of the same culturing techniques described in this book to colonize microbes in your college laboratory. We provide step-by-step instructions on how to do this.

You would find it difficult to live without the aid of microbes. For example, living inside your intestines are colonies of microorganisms. Just this thought is enough to make your skin crawl. As frightful as this thought might be, however, these microbes actually assist your body in digesting food. That is, you might have difficulty digesting some foods if these microbes did not exist.

Microbes in your intestines are beneficial to you as long as they remain in your intestines. However, you'll become very ill should they decide to wander into other parts of your body. Don't become too concerned—these microbes tend to stay at home unless your intestines are ruptured as a result of trauma.

By the end of this book you'll learn about the different types of microbes, how to identify them by using a microscope, and how to cultivate colonies of microbes.

A Look Inside

Microbiology can be challenging to learn unless you follow the step-by-step approach that is used in *Microbiology Demystified*. Topics are presented in an order in which many students like to learn them—starting with basic components and then gradually moving on to those that are more complex.

Each chapter follows a time-tested formula that explains topics in an easy-to-read style. You can then compare your knowledge with what you're expected to know by taking chapter tests and the final exam. There is little room for you to go adrift.

CHAPTER 1: THE WORLD OF THE MICROORGANISM

You'll begin your venture into the microscopic world of microbes by learning the fundamentals. These are the terms and concepts that all students need to understand before they can embark on more advanced topics, such as cultivating their own microbes.

In this chapter, you will be introduced to the science of microbiology with a look back in time to a period when little was known about microbes except that some of them could kill people. You'll also learn about the critical accomplishments made in microbiology that enable scientists to understand and develop cures for disease.

CHAPTER 2: THE CHEMICAL ELEMENTS OF MICROORGANISMS

Chemistry is a major factor in microbiology because microbes are made up of chemical elements. Scientists are able to destroy microbes by breaking them down into their chemical elements and then disposing of those elements.

Before you can understand how this process works, you must be familiar with the chemical principles related to microbiology. You'll learn about these chemical principles in this chapter.

CHAPTER 3: OBSERVING MICROORGANISMS

"Wash the germs from your hands!" That was the cry of every mom who knew that hand washing is the best way to prevent sickness. Most kids balked at hand washing simply because they couldn't see the germs on their hands.

We'll show you how to see germs and other microbes in this chapter by using a microscope. You'll learn everything you need to know to bring microbes into clear focus so you can see with a microscope what you can't see with the naked eye.

CHAPTER 4: PROKARYOTIC CELLS AND EUKARYOTIC CELLS

It is time to get down and personal with two common microbe cells. These are prokaryotic cells and eukaryotic cells. These names are probably unfamiliar to you, but they won't be by the time you're finished reading this chapter.

Prokaryotic cells are bacteria cells and eukaryotic cells are cells of animals, plants, algae, fungi, and protozoa. Each carries out the six life processes that all living things have in common. In this chapter you'll learn about how prokaryotes and eukaryotes carry out these life processes.

CHAPTER 5: THE CHEMICAL METABOLISM

"It's my slow metabolism! That's why I can't shed a few pounds." This is a great excuse for being unable to lose weight, but the reason our metabolisms are slow is because we tend not to exercise enough.

In this chapter, you'll learn about the biochemical reactions that change food into energy—collectively called metabolism—and how the cell is able to convert nutrients into energy.

CHAPTER 6: MICROBIAL GROWTH AND CONTROLLING MICROBIAL GROWTH

You and microbes need nutrients to grow—chemical nutrients such as carbon, hydrogen, nitrogen, and oxygen. However, not all microbes need the same chemical nutrients. For example, some require oxygen while others can thrive in an oxygen-free environment.

You'll learn in this chapter how to classify microbes by the chemical nutrients they need to survive. You'll also learn how to use this knowledge to grow microbes and control their growth in the laboratory.

CHAPTER 7: MICROBIAL GENETICS

Just like us, microbes inherit genetic traits from their species' previous generations. Genetic traits are instructions on how to everything to stay alive. Some instructions are passed along to the next generation while other instructions are not.

In this chapter you'll learn how microorganisms inherit genetic traits from previous generations of microorganisms. Some of these traits show them how to identify and process food, how to excrete waste products, and how to reproduce.

CHAPTER 8: RECOMBINANT DNA TECHNOLOGY

Who we are and what we are going to be is programmed into our genes. The same is true for microbes. This genetic information is encoded into DNA by the linking of nucleic acids in a specific sequence.

Genetic information can be reordered in a process called genetic engineering. You'll learn about genetic engineering and how to recombinant DNA using DNA technology in this chapter.

CHAPTER 9: CLASSIFICATION OF MICROORGANISMS

There are thousands of microbes and no two are identical, but many have similar characteristics. Microbiologists have spent years carefully observing microbes and organizing them into groups by their similarities.

You'll learn how microbes are classified in this chapter, which enables you to efficiently identify microbes that you see under a microscope.

CHAPTER 10: THE PROKARYOTES: DOMAINS ARCHAEA AND BACTERIA

Bacteria are one of the most common microbes that you encounter. Some bacteria cause disease and other bacteria help you live by aiding in digestion. There are many different kinds of bacteria; however, all bacteria can be grouped into four divisions based on the characteristics of their cell walls.

Each division is further divided into sections based on other characteristics, such as oxygen requirements, motility, shape, and Gram-stain reaction. In this chapter, you'll learn how to use these divisions and sections to identify bacteria.

CHAPTER 11: THE EUKARYOTES: FUNGI, ALGAE, PROTOZOA, AND HELMINTHS

In this chapter you'll take a close look at the kingdoms of fungi, protista, and animalia. These are microbes that are commonly known as fungi, algae, protozoa, and helminths.

Eukaryotes are a type of microbe and are different from bacteria and viruses. However, they, too, are beneficial to us. They supply food, remove waste, and cure disease (in the form of antibiotics). And as bacteria, some eukaryotes also cause disease.

CHAPTER 12: VIRUSES, VIROIDS, AND PRIONS

Probably one of the most feared microbes is a virus because often there is little or nothing that can be done to kill it. Once you're infected, you can treat the symptoms, such as a runny nose and watery eyes, but otherwise you must let the virus run its course.

Did you ever wonder why this is case? If so, then read this chapter for the answer and learn what a virus is, how viruses live, and which diseases they cause.

CHAPTER 13: EPIDEMIOLOGY AND DISEASE

It's flu season and you can only hope that you don't become infected—otherwise you'll have ten days of chills, sneezing, and isolation. No one will want to come close to you for fear of catching the flu.

In this chapter, you'll learn about diseases like the flu and how diseases are spread. You'll also learn how to take simple precautions to control and prevent the spread of diseases.

CHAPTER 14: IMMUNITY

Inside your body there is a war going on. An army of B cells, T cells, natural killer cells, and other parts of your immune system are on the defense. These cells seek microbes to rip apart before any of them can give you a runny nose, cough, or that dreaded feverish feeling.

The immune system is your body's defense mechanism: Its "soldiers" surround, neutralize, and destroy foreign invaders before they can do harm. In this chapter you'll learn about your immune system and how it gives you daily protection against invading microbes.

CHAPTER 15: VACCINES AND DIAGNOSING DISEASES

Think about this: Each year millions of people pay their doctor to inject them with the flu virus. On the surface that may not make sense, but after reading this chapter you'll find that it makes perfect sense because this injection is actually a vaccination against the flu.

A vaccine prevents you from catching a certain disease because it has elements of that disease, triggering your body to create antibodies to the disease. You'll learn about vaccines and antibodies in this chapter.

CHAPTER 16: ANTIMICROBIAL DRUGS

"Doc, give me a pill to knock out whatever is causing me to be sick!" All of us say this whenever we come down with an illness. All we want is a magic pill that makes us feel better. Sometimes that magic pill—or injection—contains a microbe that seeks out and destroys pathogenic microbes, which are disease-causing microbes.

In this chapter you'll learn about antimicrobial drugs that are given as chemotherapy to cure disease. These antimicrobial drugs contain microbes that kill other microbes.

ACKNOWLEDGMENTS

I would like to acknowledge my mentor, colleague, and friend Professor Robert Highley for all his encouragement and technical support in the production of this book; my friend Ms. Joan Sisto for all her hours of computer work; and my friend and coauthor Professor Jim Keogh for asking me to help write this book. Thank you.

DR. TOM BETSY

Professor Robert Highley has done a magnificent job as technical editor on this project. His diligence and attention to detail has made *Microbiology Demystified* a rewarding addition to every microbiology student's library.

JIM KEOGH

MICROBIOLOGY DEMYSTIFIED

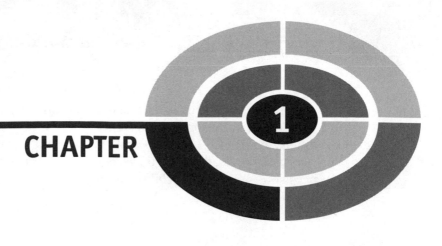

CHAPTER 1

The World of the Microorganism

Microbiology is the study of **microorganisms**, which are tiny organisms that live around us and inside our body. An *organism* is a living thing that ingests and breaks down food for energy and nutrients, excretes undigested food as waste, and is capable of reproduction. You are an organism and so are dogs, cats, insects, and other creatures that you see daily.

A microorganism is simply a very, very small organism that you cannot see with your naked eye, but you sure feel its effect whenever your eyes fill with water and mucus flows like an open faucet from your nose. You call it a head cold. Actually, you are under siege by an army of microorganisms attacking membranes inside your body. Watery eyes and a runny nose are ways that you fight microorganisms by flushing them out of your body.

Microorganisms are a key component of biological warfare along with chemicals that disrupt homeostasis. The anthrax attack that followed the 9/11 terrorist attacks clearly illustrated how a dusting of anthrax in an envelope can be lethal to people in an office building. **Anthrax** is a disease caused by the microorganism *Bacillus anthracis,* a bacterium that forms endospores and infiltrates the body by ingestion, skin contact, and inhalation. An *endospore* is a bacterium in a dormant state that forms within a cell.

Fortunately, incidents of biological attacks using microorganisms have been infrequent. However, there are thousands of microorganisms all around us that can be just as deadly and debilitating as the microorganisms used in warfare throughout history.

Types of Microorganisms

UNFRIENDLY MICROORGANISMS

An infection is caused by the infiltration of a disease-causing microorganism known as a *pathogenic microorganism*. Some pathogenic microorganisms infect humans, but not other animals and plants. Some pathogenic microorganisms that infect animals or plants also infect humans.

Pathogenic microorganisms make headlines and play an important role in history. Legendary gunfighter John "Doc" Holliday is famous for his escapades in the Wild West. He dodged countless bullets, showing that he was the best of the best when it came to gun fighting. Yet *Mycobacterium tuberculosis* took down Doc Holliday quietly, without firing a shot. *Mycobacterium tuberculosis* is the bacterium that causes tuberculosis (Fig. 1-1). This bacterium affects the lung tissue when droplets of respiratory secretions or particles of dry sputum from a person who is infected with the disease are inhaled by an uninfected person.

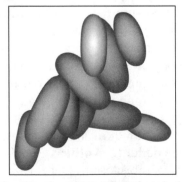

Fig. 1-1. *Mycobacterium tuberculosis* is the bacterium that causes tuberculosis.

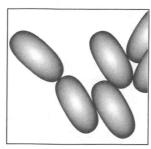

Fig. 1-2. *Yersinia pestis* is the microorganism that caused the Black Plague.

Yersinia pestis nearly conquered Europe in the fourteenth century with the help of the flea. *Yersinia pestis* is the microorganism that caused the Black Plague (Fig. 1-2) and killed more than 25 million Europeans. You might say that *Yersinia pestis* launched a sneak attack. First, it infected fleas that were carried into populated areas on the backs of rats. Rodents traveled on ships and then over land in search of food. Fleas jumped from rodents and bit people, transmitting the *Yersinia pestis* microorganism into the person's blood stream.

In an effort to prevent the spread of *Yersinia pestis*, sailors entering Sicily's seaports had to wait 40 days before leaving the ship. This gave time for sailors to exhibit the symptoms of the Black Plague if the *Yersinia pestis* microorganism had infected them. Sicilians called this *quarantenaria*. Today we know it as quarantine. Sailors who did not exhibit these symptoms were not infected and free to disembark.

Campers and travelers sometimes become acquainted with *Giardia lamblia*, *Escherichia coli*, or *Entameba histolytica* whenever they visit tropical countries. Travelers who become infected typically do not die but come down with a bad case of diarrhea.

FRIENDLY MICROORGANISMS

Not all microorganisms are pathogens. In fact many microorganisms help to maintain homeostasis in our bodies and are used in the production of food and other commercial products. For example, *flora* are microorganisms found in our intestines that assist in the digestion of food and play a critical role in the formation of vitamins such as vitamin B and vitamin K. They help by breaking down large molecules into smaller ones.

What Is a Microorganism?

Microorganisms are the subject of *microbiology,* which is the branch of science that studies microorganisms. A microorganism can be one cell or a cluster of cells that can be seen only by using a microscope.

Microorganisms are organized into six fields of study: bacteriology, virology, mycology, phycology, protozoology, and parasitology.

BACTERIOLOGY

Bacteriology is the study of bacteria. *Bacteria* are prokaryotic organisms. A *prokaryotic* organism is a one-celled organism that does not have a true nucleus. Many bacteria absorb nutrients from their environment and some make their own nutrients by photosynthesis or other synthetic processes. Some bacteria can move freely in their environment while others are stationary. Bacteria occupy space on land and can live in an aquatic environment and in decaying matter. They can even cause disease. *Bacillus anthracis* is a good example. It is the bacterium that causes anthrax.

VIROLOGY

Virology is the study of viruses. A *virus* is a submicroscopic, parasitic, acellular entity composed of a nucleic acid core surrounded by a protein coat. *Parasitic acellular* means that a virus receives food and shelter from another organism and is not divided into cells. An example of a virus is the *varicella-zoster* virus (Fig. 1-3), which is the virus that causes chickenpox in humans.

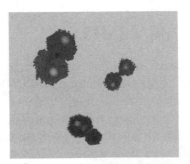

Fig. 1-3.   The *varicella-zoster* virus causes chickenpox.

MYCOLOGY

Mycology is the study of fungi. A fungus is a **eukaryotic** organism, often micro-scopic, that absorbs nutrients from its external environment. Fungi are not pho-tosynthetic. A **eukaryotic microorganism** is a microorganism whose cells have a nucleus, cytoplasm and organelles. These include yeasts and some molds. Tinea pedis, better known as athlete's foot, is caused by a fungus.

PHYCOLOGY

Phycology is the study of algae. *Algae* are *eukaryotic photosynthetic* organisms that transform sunlight into nutrients using photosynthesis. A *eukaryotic photo-synthetic* microorganism is a microorganism whose cells have a nucleus, nuclear envelope, cytoplasm, and organelles and that is able to carry out photosynthesis.

PROTOZOOLOGY

Protozoology is the study of *protozoa*, animal-like single-cell microorganisms that can be found in aquatic environments. Many obtain their food by engulfing or ingesting smaller organisms. Protozoa are found in aquatic and terrestrial environments. An example is *Amoeba proteus*.

PARASITOLOGY

Parasitology is the study of parasites. A *parasite* is an organism that lives at the expense of another organism or host. Parasites that cause disease are called *pathogens*. Examples of parasites are bacteria, viruses, protozoa, and many ani-mals such as worms, flatworms, and arthropods (insects).

What's in a Name: Naming and Classifying

Carl Linnaeus developed the system for naming organisms in 1735. This system is referred to as *binominal nomenclature*. Each organism is assigned two latin-ized names because Latin or Greek was the traditional language used by schol-ars. The first name is called the *genus*. The second name is called the *specific*

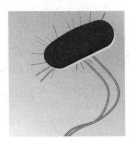

Fig. 1-4. *E. coli* is a bacterium that lives in the colon.

epithet, which is the name of the species, and is not capitalized. The genus and the epithet appear underlined or italicized.

The name itself describes the organism. For example, *Staphylococcus aureus* is a very common bacterium. *Staphylococcus* is the genus and *aureus* is the epithet. In this case, the genus describes the appearance of the cells. *Staphylo* means a clustered arrangement of the cells and *coccus* signifies that the cells are spheres. In other words, this means a cluster of sphere-like cells. Aureus is the Latin word for golden, which means that the cluster of sphere-like cells has a golden hue.

Sometimes an organism is named for a researcher, as is the case with *Escherichia coli* (Fig. 1-4), better known as *E. coli*. The genus is *Escherichia*, which is named for Theodor Escherich, a leading microbiologist. The epithet is *coli*, which implies that the bacterium lives in the colon (large intestine).

Organisms were classified into either the animal kingdom or the plant kingdom before the scientific community discovered microorganisms in the seventeenth century. It was at that time when scientists realized that this classification system was no longer valid.

Carl Woese developed a new classification system that arranged organisms according to their molecular characteristics and then cellular characteristics. However, it wasn't until 1978 when scientists could agree on the new system for classifying organisms, and it took 12 years after this agreement before the new system was published.

Woese devised three classification groups called *domains*. A domain is larger than a kingdom. These are:

Domains

- Eubacteria: Bacteria that have peptidoglycan cell walls. (Peptidoglycan is the molecular structure of the cell walls of eubacteria which consists of N-acetylglucosamine, N-acetylmuramic acid, tetrapeptide, side chain and murein.)

- Archaea: Prokaryotes that do not have peptidoglycan cell walls.
- Eucarya: Organisms from the following kingdoms:

Kingdoms
- Protista (*Note*: This is in the process of changing.)—algae, protozoa, slime molds.
- Fungi—one-celled yeasts, multicellular molds, and mushrooms.
- Plantae—moss, conifers, ferns, flowering plants, algae.
- Animalia—insects, worms, sponges, and vertebrates.

How Small Is a Microorganism?

Microorganisms are measured using the metric system, which is shown in Table 1.1 In order to give you some idea of the size of a microorganism, let's compare a microorganism to things that are familiar to you.

German shepherd	1 meter
Human gamete (egg) from a female ovary	1 millimeter
A human red blood cell	100 micrometers
A typical bacterium cell	10 micrometers
A virus	10 nanometers
An atom	0.1 nanometer

Your Body Fights Back

Immunology is the study of how an organism specifically defends itself against infection by microorganisms. When a microorganism such as the bacterium *Streptococcus pyogenes*, which can cause strep throat, invades your body, white blood cells engulf the bacterial cells and digest it in an immune response called phagocytosis. *Phagocytosis* is the ability of a cell to engulf and digest solid materials by the use of *pseudopods* or "false feet."

Phagocytosis was discovered in 1880 by Russian zoologist Elie Metchnikoff, who was one of the first scientists to study immunology. Metchnikoff studied the body's defense against disease-causing agents and invading microorganisms. He

Table 1-1.   Quantity and Length: Metric and English Equivalents

Unit	Fraction of Standard	English Equivalent
meter (m)		3.28 feet
centimeter (cm)	$0.01 \text{ m} = 10^{-2}$	0.39 inch
millimeter (mm)	$0.001 \text{ m} = 10^{-3}$	0.039 inch
micrometer (μm)	$0.000001 \text{ m} = 10^{-6}$	0.000039 inch
nanometer (nm)	$0.000000001 \text{ m} = 10^{-9}$	0.000000039 inch

discovered that leukocytes (white blood cells) defended the body by engulfing and eating the invading microorganism.

DRUGS: SEND IN THE CAVALRY

Invading microorganisms activate your body's immune system. It is at this point when you experience a fever and feel sick. In an effort to help your immune system, physicians prescribe drugs called *antibiotics* that contain one or more antimicrobial agents that combat bacteria. An antimicrobial agent is a substance that specifically inhibits and destroys the attacking microorganism.

One of the most commonly used antimicrobial agent is penicillin. *Penicillin* is made from Penicillium, which is a mold that secretes materials that interfere with the synthesis of the cell walls of bacteria causing "lysis," or destruction of the cell wall, and kills the invading microorganism.

IMMUNITY: PREVENTING A MICROORGANISM ATTACK

Our bodies have a wide range of body responses in the fight against pathogens. These responses are referred to as *nonspecific resistance*. *Resistance* is the ability of the body to ward off disease. The lack of resistance is called *susceptibility*. When your immune system is compromised you become susceptible to pathogens invading your body where they divide into colonies causing disease, making you sick.

Generally, your first line of defense is to use mechanical and chemical means to prevent a pathogen from entering your body. Skin is the primary mechanical means to fight pathogens; it acts as a barrier between the pathogen and the internal structures of your body. Mucous membranes are another mechanical barrier;

they move the pathogen using tears and saliva (to flush it) and mucus cilia in the respiratory track to physically move it. Urination, defecation, and vomiting are other mechanical means to combat a pathogen by forcefully removing the pathogen from your body.

Chemical means attack a pathogen by changing its pH properties. Skin sebum is an important defense. Sebum is a thick substance secreted by the sebaceous glands; it consists of lipids and cellular debris that have a low pH, enabling it to chemically destroy a pathogen.

Sweat contains the enzyme lysozyme, which attacks the cell walls of bacteria. Hyaluronic acid found in areolar connective tissue sets up a chemical barrier that restricts a pathogen to a localized area of the body. Likewise, gastric juice and vaginal secretions have a low pH that is a natural barrier to many kinds of pathogens.

History of the Microscope

Diseases are less baffling today than they were centuries ago, when scientists and physicians were clueless as to what causes disease. Imagine for a moment that a close relative had taken ill. One day she was well and the next day she was sick for no apparent reason. Soon she was dead if her body couldn't fight the illness. You couldn't see whatever attacked her—and neither could the doctor.

ZACHARIAS JANSSEN

In 1590, Zacharias Janssen developed the first compound microscope in Middleburg, Holland. Janssen's microscope consisted of three tubes. One tube served as the outer casing and contained the other two tubes. At either ends of the inner tubes were lenses used for magnification. Janssen's design enabled scientists to adjust the magnification by sliding the inner tubes. This enabled scientists to enlarge the image of a specimen three and nine times the specimen's actual size.

ROBERT HOOKE

In 1665, Robert Hooke, an English scientist, popularized the use of the compound microscope when he placed the lenses over slices of cork and viewed little boxes that he called cells. It was his discovery that led to the development of cell theory in the nineteenth century by Mathias Schleiden, Theodor Schwann, and Rudolf Virchow. *Cell theory* states that all living things are composed of cells.

ANTONI VAN LEEUWENHOEK

Hooke's experiments with a crude microscope inspired Antoni van Leeuwenhoek to further explore the micro world. Van Leeuwenhoek, an amateur lens grinder, improved Hooke's microscope by grinding lenses to achieve magnification. His microscope required one lens. With his improvement, van Leeuwenhoek became the first person to view a living microorganism, which he called *Animalcules*.

This discovery took place during the 1600s, when scientists believed that organisms generated spontaneously and did not come from another organism. This sounds preposterous today; however, back then scientists were just learning that a cell was the basic component of an organism.

How Do Organisms Appear?

FRANCESCO REDI

Italian physician Francesco Redi developed an experiment that demonstrated that an organism did not spontaneously appear. He filled jars with rotting meat. Some jars he sealed and others he left opened. Those that were open eventually contained maggots, which is the larval stage of the fly. The other jars did not contain maggots because flies could not enter the jar to lay eggs on the rotting meat.

His critics stated that air was the ingredient required for spontaneous generation of an organism. Air was absent from the sealed jar and therefore no spontaneous generation could occur, they said (Fig. 1-5). Redi repeated the experiment except this time he placed a screen over the opened jars. This prevented flies from entering the jar. There weren't any maggots on the rotting meat.

Until that time scientists did not have a clue about how to fight disease. However, Redi's discovery gave scientists an idea. They used Redi's findings to conclude that killing the microorganism that caused a disease could prevent the disease from occurring. A new microorganism could only be generated by

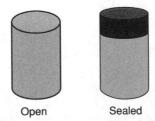

Open   Sealed

Fig. 1-5.   No spontaneous generation occurred in the sealed jar.

the reproduction of another microorganism. Kill the microorganism and you won't have new microorganisms, the theory went—you could stop the spread of the disease. Scientists called this the *Theory of Biogenesis*. The Theory of Biogenesis states that a living cell is generated from another living cell.

LOUIS PASTEUR

Although the Theory of Biogenesis disproved spontaneous generation, spontaneous generation was hotly debated among the scientific community until 1861 when Louis Pasteur, a French scientist, resolved the issue once and for all. Pasteur showed that microorganisms were in the air. He proved that sterilized medical instruments became contaminated once they were exposed to the air.

Pasteur came to this conclusion by boiling beef broth in several short-necked flasks. Some flasks were left open to cool. Other flasks were sealed after boiling. The opened flasks became contaminated with microorganisms while no microorganisms appeared in the closed flasks. Pasteur concluded that airborne microorganisms had contaminated the opened flasks.

In a follow-up experiment, Pasteur placed beef broth in an open long-necked flask. The neck was bent into an *S*-shape. Again he boiled the beef broth and let it cool. The *S*-shaped neck trapped the airborne microorganisms (see Fig. 1-6).

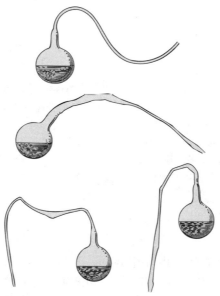

Fig. 1-6.   Pasteur placed beef broth into a long-necked flask, then bent the neck into an S-shape.

The beef broth remained uncontaminated even after months of being exposed to the air. The very same flask containing the original beef broth exists today in Pasteur Institute in Paris and still shows no sign of contamination. Pasteur's experiments validated that microorganisms are not spontaneously generated.

Based on Pasteur's findings, a concerted effort was launched to improve sterilization techniques to prevent microorganisms from reproducing. *Pasteurization,* one of the best-known sterilization techniques, was developed and named for Pasteur. Pasteurization kills harmful microorganisms in milk, alcoholic beverages, and other foods and drinks by heating it enough to kill most bacteria that cause spoilage.

JOHN TYNDALL AND FERDINAND COHN

The work of John Tyndall and Ferdinand Cohn in the late 1800s led to one of the most important discoveries in sterilization. They learned that some microorganisms are resistant to certain sterilization techniques. Until their discovery, scientists had assumed that no microorganism could survive boiling water, which became a widely accepted method of sterilization. This was wrong. Some thermophiles resisted heat and could survive a bath in boiling water. This meant that there was not one magic bullet that killed all harmful microorganisms.

Germ Theory

Until the late 1700s, not much was really known about diseases except their impact. It seemed that anyone who came in contact with an infected person contracted the disease. A disease that is spread by being exposed to infection is called a *contagious disease*. The unknown agent that causes the disease is called a *contagion*. Today we know that a contagion is a microorganism, but in the 1700s many found it hard to believe something so small could cause such devastation.

ROBERT KOCH

Opinions changed dramatically following Robert Koch's study of anthrax in the late 1800s. Koch noticed a pattern developing: Anyone who worked with or ingested animals that were infected with anthrax contracted the disease. In fact,

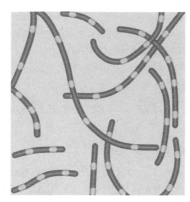

Fig. 1-7. *Bacillus anthracis* rapidly multiplies in the active state and becomes infectious.

people who simply inhaled the air around an infected animal were likely to inhale the anthrax bacterium spores and come down with the disease. Koch's investigations into anthrax led him to discover how microorganisms work.

Anthrax is caused by *Bacillus anthracis* (Fig. 1-7), which is a bacterial type of microorganism consisting of one cell. *Bacillus anthracis,* whether in a *dormant* or an *active state,* is called a *spore*. A spore is not infectious. However, under the right conditions, the *Bacillus anthracis* spores germinate and enter the active state and rapidly multiply and become infectious.

The question that Koch raised is: Would taking active *Bacillus anthracis* from one animal and injecting it into a healthy animal cause the healthy animal to come down with anthrax? If so, then he could prove that a microorganism was actually the cause of disease.

Bacillus anthracis was present in the blood of infected animals, so Koch removed a small amount of blood and injected it into a healthy animal. The animal came down with anthrax. He repeated the experiment by removing a small amount of blood from the newly infected animal and gave it to another healthy animal. It, too, came down with anthrax.

Koch expanded his experiment by cultivating *Bacillus anthracis* on a slice of potato. He then exposed the potato to the right blend of air, nutrients, and temperature. Koch took a small sample of his homegrown *Bacillus anthracis* and injected it into a healthy animal. The animal came down with anthrax.

Based on his findings, Koch developed the Germ Theory. The *Germ Theory* states that a disease-causing microorganism should be present in animals infected by the disease and not in healthy animals. The microorganism can be cultivated away from the animal and used to inoculate a healthy animal. The healthy animal should then come down with the disease. Samples of a microorganism taken from several infected animals are the same as the original microorganism from the first infected animal.

Four steps used by Koch to study microorganisms are referred to as Koch's Postulates. *Koch's Postulates* state:

1. The microorganism must be present in the diseased animal and not present in the healthy animal.

2. Cultivate the microorganism away from the animal in a pure culture.

3. Symptoms of the disease should appear in the healthy animal after the healthy animal is inoculated with the culture of the microorganism.

4. Isolate the microorganism from the newly infected animal and culture it in the laboratory. The new culture should be the same as the microorganism that was cultivated from the original diseased animal.

Koch's work with anthrax also developed techniques for growing a culture of microorganisms. He eventually used a gelatin surface to cultivate microorganisms. Gelatin inhibited the movement of microorganisms. As microorganisms reproduced, they remained together, forming a colony that made them visible without a microscope. The reproduction of microorganisms is called *colonizing*. The gelatin was replaced with agar that is derived from seaweed and still used today. Richard Petri improved on Koch's cultivating technique by placing the agar in a specially designed disk that was to later be called the Petri dish (Fig. 1-8). It, too, is still used today.

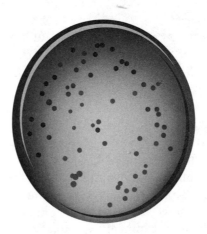

Fig. 1-8. A Petri dish is used to grow a culture of microorganisms.

Vaccination

The variola microorganism was one of the most feared villains in the late 1700s. The variola virus causes smallpox. If variola didn't kill you, it caused pus-filled blisters that left deep scars that pitted nearly every part of your body. Cows were also susceptible to a variation of variola called *cowpox*. Milkmaids who tended to infected cows contracted cowpox and exhibited immunity to the smallpox virus.

EDWARD JENNER

Edward Jenner, an English physician, discovered something very interesting about both smallpox and cowpox in 1796. Those who survived smallpox never contracted smallpox again, even when they were later exposed to someone who was infected with smallpox. Milkmaids who contracted cowpox never caught smallpox even though they were exposed to smallpox.

Jenner had an idea. He took scrapings from a cowpox blister found on a milkmaid and, using a needle scratched the scrapings into the arm of James Phipps, an 8-year-old. Phipps became slightly ill when the scratch turned bumpy. Phipps recovered and was then exposed to smallpox. He did not contract smallpox because his immune system developed antibodies that could fight off variola and vaccinia.

Jenner's experiment discovered how to use our body's own defense mechanism to prevent disease by *inoculating* a healthy person with a tiny amount of the disease-causing microorganism. Jenner called this a *vaccination*, which is an extension of the Latin word *vacca* (cow). The person who received the vaccination became *immune* to the disease-causing microorganism.

ELIE METCHNIKOFF

Elie Metchnikoff, a nineteenth-century Russian zoologist, was interested by Jenner's work with vaccinations. Metchnikoff wanted to learn how our bodies react to vaccinations by exploring our body's immune system. He discovered that white blood cells (*leukocytes*) engulf and digest microorganisms that invade the body. He called these cells *phagocytes*, which means "cell eating." Metchnikoff was one of the first scientists to study the new area of biology called *immunology*, the study of the immune system.

Killing the Microorganism

Great strides were made during the late 1800s in the development of antiseptic techniques. It began with a report by Hungarian physician Ignaz Semmelweis on a dramatic decline in childbirth fever when physicians used antiseptic techniques when delivering babies. Infections become preventable through the use of antiseptic techniques.

JOSEPH LISTER

Joseph Lister, an English surgeon, developed one of the most notable antiseptic techniques. During surgery he sprayed carbolic acid over the patient and then bandaged the patient's wound with carbolic acid–soaked bandages. Infection following surgery dramatically dropped when compared with surgery performed without spraying carbolic acid. *Carbolic acid,* also known as *phenol* was one of the first surgical antiseptics.

PAUL EHRLICH

Antiseptics prevented microorganisms from infecting a person, but scientists still needed a way to kill microorganisms after they infected the body. Scientists needed a magic bullet that cured diseases. At the turned of the nineteenth century, Paul Ehrlich, a German chemist, discovered the magic bullet. Ehrlich blended chemical elements into a concoction that, when inserted into an infected area, killed microorganisms without affecting the patient. Today we call Ehrlich's concoction a drug. Ehrlich's innovation has led to chemotherapy using synthetic drugs that are produced by chemical synthesis.

ALEXANDER FLEMING

Scientists from all over set out to use Ehrlich's findings to find drugs that could make infected patients well again. One of the most striking breakthroughs came in 1929 when Alexander Fleming discovered *Penicillin notatum,* the organism that synthesizes penicillin. *Penicillium notatum* is a fungus that kills the *Staphylococcus aureus* microorganism (Fig. 1-9), and similar microorganisms.

Fig. 1-9. *Penicillium notatum* is a fungus
that kills the *staphylococcus aureus.*

Fleming grew cultures of Staphylococcus aureus, a bacterium, in the laboratory. He was also conducting experiments with *Penicillium notatim*, a mold. By accident the Staphylococcus aureus was contaminated with the *Penicillium notatum*, causing the Staphylococcus to stop reproducing and die. *Penicillium notatum* became one of the first antibiotics. An **antibiotic** is a substance that kills bacteria.

A summary of the scientists and their contributions can be found in Table 1-2.

Table 1-2. Scientists and Their Contributions

Year	Scientist	Contribution
1590	Zacharias Janssen	Developed the first compound microscope.
1590	Robert Hooke	Observed nonliving plant tissue of a thin slice of cork.
1668	Francesco Redi	Discovered that microorganisms did not spontaneously appear. His contribution led to the finding that killing the microorganism that caused the disease could prevent the disease.
1673	Antoni van Leeuwenhoek	Invented the single-lens microscope, grinding the microscope lens to improve magnification. First person to view a living organism.
1798	Edward Jenner	Developed vaccinations against disease-causing microorganisms.

Table 1-2.   Scientists and Their Contributions *(Continued)*

Year	Scientist	Contribution
1850s	Mathias Schleiden, Theodore Schwann, Rudolf Virchow	Developed cell theory.
1847	Ignaz Semmelweis	Reported a dramatic decline in childbirth fever after physicians used antiseptic techniques when delivering babies.
1864	Louis Pasteur	Discovered that microorganisms were everywhere, living on organisms and in nonliving things such as air. His work led to improved sterilization techniques called pasteurization. One of the founders of bacteriology.
1867	Joseph Lister	Reduced infections after surgery by spraying carbolic acid over the patient before bandaging the wound. This was the first surgical antiseptic.
1876	Robert Koch	Discovered how microorganisms spread contagious diseases by studying anthrax. Developed the Germ Theory. Developed techniques for cultivating microorganisms.
1870s	John Tyndall, Ferdinand Cohn	Discovered that some microorganisms are resistant to certain sterilization techniques. One of the founders of bacteriology.
1884	Elie Metchnikoff	Discovered that white blood cells (*leukocytes*) engulf and digest microorganisms that invade the body. Coined the word *phagocytes*. Founded the branch of science called immunology.
1887	Richard Petri	Developed the technique of placing agar into a specially designed dish to grow microorganisms, which was later called the Petri dish.
1890	Paul Ehrlich	Developed the first drug to fight disease-causing microorganisms that had already entered the body.
1928	Alexander Fleming	Discovered *Penicillium notatum,* the fungus that kills *staphylococcus aureus,* a microorganism that is a leading cause of infection.

Quiz

1. What is a microorganism?
   (a) A microorganism is a small organism that takes in and breaks down food for energy and nutrients, excretes unused food as waste, and is capable of reproduction.
   (b) A microorganism is a small organism that causes diseases only in plants.
   (c) A microorganism is a small organism that causes diseases only in animals.
   (d) A microorganism is a term that refers to a cell.

2. What is a pathogenic microorganism?
   (a) A microorganism that multiplies
   (b) A microorganism that grows in a host
   (c) A microorganism that is small
   (d) A disease-causing microorganism

3. Name the parts of this microorganism using the nomenclature system: *Mycobacterium tuberculosis*.
   (a) A bacterium is a one-cell organism that does not have a distinct nucleus.
   (b) *Mycobacterium* is the presemous and *tuberculosis* is the specific postsemous.
   (c) *Mycobacterium* is the epithet and *tuberculosis* is the specific genus.
   (d) *Mycobacterium* is the genus and *tuberculosis* is the specific epithet.

4. Why is a bacterium called a prokaryotic organism?
   (a) A bacterium is a one-cell organism that does not have a distinct nucleus.
   (b) A bacterium is a one-cell organism that has a distinct nucleus.
   (c) A bacterium is a multicell organism that does not have a distinct nucleus.
   (d) A bacterium is a multicell organism that has a distinct nucleus.

5. Why is a fungus called a eukaryotic microorganism?
   (a) Fungus has cells that have a nucleus, nuclear envelope, cytoplasm, and organelles.

    (b) Fungus has cells that have a nucleus and no nuclear envelope.

    (c) Fungus has cells that have a nucleus, nuclear envelope, cytoplasm, but no organelles.

    (d) Fungus has cells that have no nucleus, no nuclear envelope, no cytoplasm, and no organelles.

6. What is Archaea?

    (a) Archaea is a classification for organisms that have two nuclei.

    (b) Archaea is a classification for organisms that use phagocytosis.

    (c) Archaea is a classification of an organism that identifies prokaryotes that do not have peptidoglycan cell walls.

    (d) Archaea is a classification of an organism that identifies prokaryotes that have peptidoglycan cell walls.

7. What is phagocytosis?

    (a) The ability of a cell to reproduce.

    (b) The ability of a cell to move throughout a microorganism.

    (c) The ability of a cell to engulf and digest solid materials by use of pseudopods, or "false feet."

    (d) The ability of a cell to change shape.

8. What is a compound microscope?

    (a) A microscope that has one lenses.

    (b) A microscope that has two sets of lenses: an ocular lens and an eyepiece.

    (c) A microscope whose lenses are concave.

    (d) A microscope whose lenses are convex.

9. What is Germ Theory?

    (a) Germ Theory states that a disease-causing microorganism should be present in animals infected by the disease and not in healthy animals.

    (b) Germ Theory states that a disease-causing microorganism should be present in healthy animals and not in infected animals.

    (c) Germ Theory states that a disease-causing microorganism should be destroyed.

    (d) Germ Theory states that a disease-causing microorganism cannot be destroyed.

10. What is Edward Jenner's contribution to microbiology?

    (a) Edward Jenner discovered the Germ Theory.

(b) Edward Jenner discovered how to create vaccinations to trigger the body's immune system to develop antibodies that fight micro-organisms.

(c) Edward Jenner discovered the compound microscope.

(d) Edward Jenner discovered the compound nomenclature system.

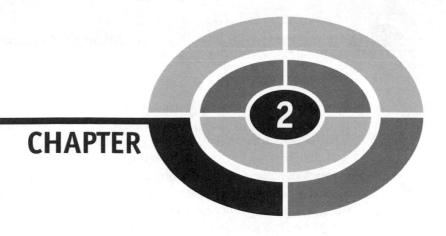

The Chemical Elements of Microorganisms

No doubt you're asking yourself what chemistry has to do with microbiology since they seem to be two different branches of science. The simple answer is that microorganisms are made up of chemicals, as is every organism—and all matter. Remember that matter is anything that occupies space and has mass.

You might say that an organism is a chemical processing plant where things are broken down into chemical elements; these chemical elements are then rearranged to form new things. You do this every time you ingest food. Food is a group of chemical compounds. The digestive process rearranges these digested chemical compounds into new substances that provide you with energy and nutrients that are necessary for you to live. Some microorganisms called *autotrophic organisms* manufacture their own food. Microorganisms that derive energy from other microorganisms (food) are called *heterotopic organisms*.

When you catch a cold or become infected by pathogenic microorganisms, your body is no longer in homeostasis. You feel rotten, but what's really happening is that the microorganism is disrupting your chemical processing plant's normal operation. Some microorganisms prevent necessary chemical processing from occurring. Other microorganisms cause your chemical processing plant to execute different processes designed to fight the microorganism attack and return your body to homeostasis—then your body is back to normal.

As you can see, chemistry is a crucial component of microbiology. It is for this reason that we begin the study of microorganisms with a close look at chemistry.

Everything Matters

Anything that takes up space and has mass is matter. The chair you're sitting on is matter. You are matter. And so are the microorganisms crawling over you and the chair. All nonliving and living things are matter because they take up space and have mass.

Matter is anything that takes up space and has mass. It is easy to envision something taking up space, but what is mass?

Mass is the amount of matter a substance or an object contains. A common misconception is that mass is the *weight* of a substance. It is true that the more there is of a substance, the more it weighs. However, weight is the force of gravity acting on mass and is calculated as weight = mass × gravity. A trip to the moon will clarify the difference between mass and weight: You have the same mass on earth as you do on the moon, but you weigh more on the earth than you do on the moon because the earth has six times the gravitational force of the moon.

Chemical Elements and the Atom

Everything including you is composed of chemical elements. A *chemical element,* sometimes simply referred to as an *element*, is a substance that cannot be broken down into simpler substances by a chemical process. All matter is a combination of chemical elements.

A chemical element is made up of atoms. An *atom* is the smallest particle of an element; it cannot be further decomposed into smaller chemical substance (Fig. 2-1). In the early 1800s, John Dalton developed the Atomic Theory, which explains the relationship between an element and an atom. The *Atomic Theory*

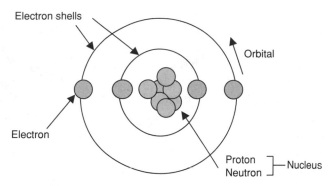

Fig. 2-1.   An atom is the smallest particle of an element.

states that an element cannot be decomposed into two or more chemical substances because the element consists of one kind of atom. The atom is also the smallest amount of matter that can enter into a chemical reaction. You'll learn about chemical reactions later in this chapter.

At the center of every atom is a *nucleus*. The nucleus does not change spontaneously unless it is unstable—making it radioactive—and does not participate in a chemical reaction. It is for this reason that the nucleus for most atoms is considered stable.

Moving around the nucleus are *electrons*. An electron is a negatively charged particle that follows a path called an *orbital*. Electrons are the parts of an atom that enter into a chemical reaction.

The nucleus is made up of protons and neutrons. A *proton* is a positively charged particle. A *neutron* is a particle that does not have a charge; it is called neutral or uncharged. The number of protons in the nucleus equals the number of electrons in an electrically stable atom. This makes the atom neutral because the number of positively charged particles (protons) offsets the number of negatively charged particles (electrons).

An element is identified by its atomic number. The *atomic number* is the number of protons in the nucleus of the atom. The *atomic mass* (also called the *atomic weight*) is slightly less than the sum of the masses of an atom's neutrons and protons. The standard for measuring atomic mass is called a *dalton*, named for John Dalton. A dalton is also known as an *atomic mass unit* (amu). For example, a neutron has an atomic mass of 1.088 daltons. A proton has an atomic mass of 1.077 daltons. An electron has an atomic mass of 0.0005 dalton.

Atoms that have the same atomic number are classified as the same chemical element because these atoms behave the same way. Therefore, a chemical element consists of one or more atoms that have the same atomic number.

Atoms of elements that have the same atomic number, but different mass numbers are called isotopes. This difference is do to a difference in number of neutrons.

Each chemical element is identified by a one or two-letter symbol that corresponds to the first letter or the first two letters in its name. For example, the symbol C is used for carbon. Some chemical elements have English names while others have Latin names. It is for this reason that symbols for some chemical elements seem strange at first glance. Take sodium, for example. You would think its symbol should be S, but that's the symbol for sulfur. The symbol for sodium is Na—the first two letters of its Latin name, natrium.

There are 92 natural chemical elements and others that scientists synthesized (created). All of these are organized into a table called the Periodic Table (see "A Dinner Table of Elements: The Periodic Table"). The six most abundant chemical elements in living things are carbon, oxygen, hydrogen, nitrogen, phosphorus and calcium. The rest are also important (see Table 2-1) and are found in trace amounts.

Table 2-1. Chemical Elements Commonly Found in All Living Things

Element	Symbol	Atomic Number	Approximate Atomic Weight
Calcium	Ca	20	40
Carbon	C	6	12
Chlorine	Cl	17	35
Hydrogen	H	1	1
Iodine	I	53	127
Iron	Fe	26	56
Magnesium	Mg	12	24
Nitrogen	N	7	14
Oxygen	O	8	16
Phosphorus	P	15	31
Potassium	K	19	39
Sodium	Na	11	23
Sulfur	S	16	32

A Dinner Table of Elements: The Periodic Table

As scientists continued to discover new chemical elements, it became apparent that there needed to be a way to place chemical elements in some kind of order. In this way, scientists can easily reference information about each chemical element.

In the 1800s Russian chemist Dimitri Mendeleev organized chemical elements into a table by their atomic weight. Chemist H. G. J. Moseley reorganized chemical elements using their atomic number rather than atomic weight. Chemical elements were placed on the table in increasing atomic number. This is referred to as the *Law of Chemical Periodicity,* and the table became known as the *Periodic Table* (Fig. 2-2).

The Periodic Table consists of seven rows, each called a *period.* Chemical elements that have the same number of electron shells are placed in the same period. Rows are divided into columns, which are identified with the Roman numerals IA through VIIIA or 1 through 18, depending on the author of the Periodic Table. Chemical elements within the same column have the same chemical properties. For example, chemical elements in column IA can easily be joined with other chemical elements. In contrast, chemical elements in column VIIIA will not join with other chemical elements.

Each chemical element is identified by its symbol on the Periodic Table and is associated with two numbers. The number on top of the chemical symbol is the atomic number. The number beneath the chemical symbol is the atomic weight.

The Glowing Tale of Isotopes

Scientists describe the decay of an isotope using half-life. The *half-life* of an isotope is the time required for half the isotope's radioactive atoms in a sample of the isotope to decay into a more stable form. The rate at which the number of atoms of an isotope disintegrates is called the isotope's *rate of decay,* which can be a matter of seconds, minutes, hours, days, or years. Ernest Rutherford coined the term half-life at the turn of the twentieth century. Rutherford discovered two kinds of radiation that he called alpha and beta. Scientists acknowledged Rutherford's important contribution by naming an element for him: rutherfordium (Rf).

Around the same time, Marie Curie along with her husband Pierre Curie discovered that atoms of the chemical element polonium (Po) and of the chemical element radium (Ra) spontaneously decayed and gave off particles. She called this process *radioactivity.*

The Periodic Table

Period	I A	II A	III A	IV A	V A	VI A	VII A	VIII A	VIII A	VIII A	I B	II B	III B	IV B	V B	VI B	VII B	VIII B
1	1 H 1.0079																	2 He 4.003
2	3 Li 6.94	4 Be 9.0121											5 B 10.81	6 C 12.011	7 N 14.006	8 O 15.999	9 F 18.998	10 Ne 20.17
3	11 Na 22.989	12 Mg 24.035											13 Al 26.981	14 Si 28.085	15 P 30.973	16 S 32.06	17 Cl 35.453	18 Ar 39.948
4	19 K 39.098	20 Ca 40.08	21 Sc 44.955	22 Ti 47.90	23 V 50.941	24 Cr 51.996	25 Mn 54.938	26 Fe 55.847	27 Co 58.933	28 Ni 58.71	29 Cu 63.546	30 Zn 65.38	31 Ga 69.735	32 Ge 72.59	33 As 74.921	34 Se 78.96	35 Br 79.904	36 Kr 83.80
5	37 Rb 85.467	38 Sr 87.62	39 Y 88.905	40 Zr 91.22	41 Nb 92.906	42 Mo 95.94	43 Tc 98.906	44 Ru 101.07	45 Rh 102.90	46 Pd 106.4	47 Ag 107.86	48 Cd 112.41	49 In 114.82	50 Sn 118.69	51 Sb 121.75	52 Te 127.60	53 I 126.90	54 Xe 131.30
6	55 Cs 132.90	56 Ba 137.33	57 La 138.90	72 Hf 178.49	73 Ta 180.94	74 W 183.85	75 Re 186.20	76 Os 190.2	77 Ir 192.22	78 Pt 195.09	79 Au 196.96	80 Hg 200.59	81 Tl 204.37	82 Pb 207.2	83 Bi 208.98	84 Po (209)	85 At (210)	86 Rn (222)
7	87 Fr (223)	88 Ra 226.02	89 Ac (227)	104 Unq (261)	105 Unp (262)	106 Unh (263)	107 Uns (262)	108 Uno (265)	109 Une (266)	110 Unn (272)								

Lanthanide Series

58 Ce 140.12	59 Pr 140.90	60 Nd 144.24	61 Pm (145)	62 Sm 150.4	63 Eu 151.96	64 Gd 157.25	65 Tb 158.92	66 Dy 162.5	67 Ho 164.93	68 Er 167.26	69 Tm 168.93	70 Yb 173.04	71 Lu 174.96

Actinide Series

90 Th 232.03	91 Pa 231.03	92 U 238.02	93 Np 237.04	94 Pu (244)	95 Am (243)	96 Cm (247)	97 Bk (247)	98 Cf (251)	99 Es (254)	100 Fm (257)	101 Md (258)	102 No (259)	103 Lr (260)

Fig. 2-2. The Periodic Table organizes chemical elements into periods and groups.

A chemical element can have multiple isotopes. Each of those isotopes has the same atomic number but different mass number. As you'll recall from earlier in this chapter, the mass number is the sum of protons and neutrons in the nucleus. Each isotope of the same chemical element has a different number of neutrons but the same number of protons.

Around They Go: Electronic Configuration

Previously in this chapter you learned that electrons of an atom move around the atom's nucleus in a pattern called an orbital. An orbital of an atom is organized into one or more energy levels around the nucleus. The lowest energy level orbital is closest to the nucleus. The highest energy level is in the outermost orbital. The outermost orbital is called the *valence shell*. Each orbital holds a maximum number of electrons.

An atom with completely filled shells is called an inert atom and is *chemically stable*. An inert atom tends not to react with other atoms. However, an atom that has an incomplete set of electrons in its valence shell is *chemically unstable* and tends to react with other atoms in an effort to become stable. Atoms want to be stable, so they either empty or fill their valence shell.

If an atom's valance shell is not filled, it is considered unstable. In order to become stable the atom must undergo a chemical reaction to acquire one or more electrons from another atom, give up one or more electrons to another atom, or share one or more electrons with another atom.

A *chemical reaction* is a chemical change in which substances called *reactants* change into substances called *products* by rearrangement, combination, or separation of elements. Chemical reactions occur naturally, sometimes taking a relatively long time to complete. A *catalyst* can be used to speed up a chemical reaction. Enzymes are chemical substances that act like catalysts to increase the rate of reaction, without changing the products of the reaction or by being consumed in the reaction. A catalyst remains unaffected by the chemical reaction and does not affect the result of the reaction. It simply speeds up the reaction.

Before James There Was Bond . . . Chemical Bond

An atom stabilizes by bonding with another atom in order to fill out its outer set of electrons in its valence shell. When two atoms of the same chemical element

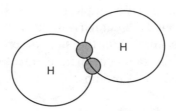

Fig. 2-3. Hydrogen becomes chemically stable by sharing a valence
electron with another hydrogen atom.

bond together they form a *diatonic molecule.* When two atoms of different chemical elements bond, they form a chemical *compound.*

Atoms are held together because there is an electrostatic attractive force
between the two atoms. Energy is required for the chemical reaction to bond
atoms. This energy becomes potential chemical energy that is stored in a molecule or chemical compound.

For example, combining two atoms of hydrogen forms a hydrogen molecule,
H_2 (Fig. 2-3). Combing a hydrogen molecule consisting of two atoms with one
oxygen atom forms the compound we know as water, H_2O (Fig. 2-4).

Bonds are formed in two ways:

- Gain or lose an electron from the valence shell; called an ionic attraction.
- Share one or more electrons in the valence shell; called a covalent bond.

In reality, atoms bond together using a range of ionic and covalence bonds.

There are four kinds of chemical bonds:

- *Ionic bond.* Transfer electrons from one atom to another atom. An atom
  becomes unbalanced when it gains or loses an electron. An atom that gains
  an electron becomes negatively charged. An atom that loses an electron

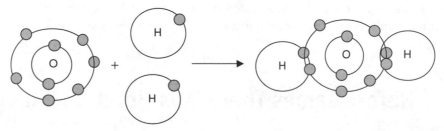

Fig. 2-4. Water is a compound consisting of two hydrogen atoms and one oxygen atom.

becomes positively charged. The atom is *oxidized.* An atom that is involved in this exchange is called an *ion.* The atom that gives up an electron is called a *cation.* A cation is positively charged. The atom that receives an electron is called an *anion,* which is negatively charged. The reaction that creates table salt from sodium and chlorine causes an ionic bond between these atoms (Fig. 2-5).

- *Covalent bond.* Atoms share electrons in their valence shell (Fig. 2-5). The shared electron orbits the nucleus of both atoms. A covalent bond is the strongest bond and the most commonly found in organisms. There are three kinds of covalent bonds: *single, double,* and *triple.* These names reflect the number of electrons that are shared between the two atoms that form the bond. Atoms that share electrons *equally* form **nonpolar covalent bond.** Atoms that share electrons *unequally* form **polar covalent bond.**

- *Coordinate covalent bond.* A bond is formed when electrons of the shared pair come from the same atom.

- *Hydrogen bond.* A hydrogen bond forms a weak (5% the strength of a covalent bond), temporary bond that serves as a bridge between either different molecules or portions of the same molecule. For example, two water molecules are physically combined using a hydrogen bond.

Decoding Chemical Shorthand

Over the years chemists have developed a way of describing atoms, chemical elements, and reactions so they can convey ideas to each other. Table 2-2 shows commonly used chemical notations that you'll need to know when learning about microbiology.

I Just Want to See Your Reaction

The process of bonding together atoms and separating atoms that are already bonded together is called a *chemical reaction.* A chemical reaction causes a change in the properties of atoms or to a collection of atoms, but the atoms remain unchanged because of a change in the electron configuration. For example,

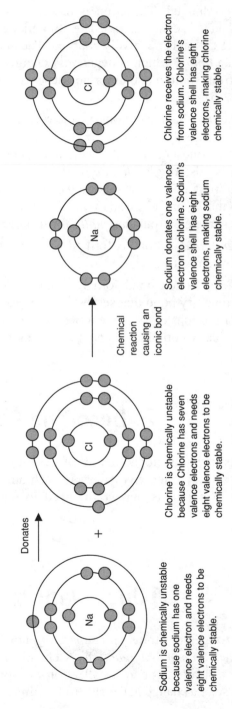

Sodium is chemically unstable because sodium has one valence electron and needs eight valence electrons to be chemically stable.

Donates

+

Chlorine is chemically unstable because Chlorine has seven valence electrons and needs eight valence electrons to be chemically stable.

Chemical reaction causing an iconic bond

Sodium donates one valence electron to chlorine. Sodium's valence shell has eight electrons, making sodium chemically stable.

Chlorine receives the electron from sodium. Chlorine's valence shell has eight electrons, making chlorine chemically stable.

Fig. 2-5. Sodium donates one valence electron to chlorine in a chemical reaction that forms the compound known as salt.

Table 2-2. Commonly Used Chemical Notations

Notation	Description
Na^+	The plus superscript indicates a positive ion.
Cl^-	The negative superscript indicates a negative ion.
$Na^+ + Cl^- \rightarrow NaCl$	The plus sign indicates synthesizing (combining) two particles. The right arrow indicates that a chemical reaction occurs towards the product.
$NaCl \rightarrow Na^+ + Cl^-$	Decomposing (breaking up) a molecule or chemical compound.
$NaOH + HCl \rightarrow NaCl + H_2O$	Exchange reaction where a chemical compound is decomposed into its chemical elements and those chemical elements are synthesized into a new compound. Here, sodium hydroxide (NaOH) and hydrochloric acid (HCl) form salt (NaCL) and water (H_2O).
$Na^+ + Cl^- \rightleftarrows NaCl$	Reversible reaction is noted with a right arrow over a left arrow.
$C - C$	Single covalent bond.
$C = C$	Double covalent bond.
$C \equiv C$	Triple covalent bond.
H_2O	A subscript following a chemical symbol indicates the number of atoms (two hydrogen atoms). If no subscript is used, then it is implied there is one atom (here, one oxygen atom).

a chemical reaction occurs when a sodium atom is combined with a chlorine atom; the property of the resulting chemical compound is table salt. If the sodium chloride (table salt) compound were broken down into its chemical elements, you would see that the atoms of sodium and chlorine remain unchanged.

Theoretically a chemical reaction can be reversed if the conditions are optimal. A chemical reaction that is reversible is called a *reversible reaction*. (see Fig. 2-6)

In practical use, same reactions can do this much easier than others. Some of these reversible reactions occur due to the instability of the reactants and products, while others will only reverse under special conditions. Examples of special conditions could be the presence of water or the application of heat.

The type of reaction that occurs can further describe a chemical reaction. There are three types of chemical reactions:

- *Synthesis* reaction: Two or more atoms, ions, or molecules are bound to form a larger molecule. A synthesis reaction combines substances called *reactants* to form a new molecule, which is called a product. A *reactant* is a substance that reacts in a reaction and the *product* is the result of a reaction. In $Na^+ + Cl^- \rightarrow NaCl$, sodium and chlorine are reactants and sodium chloride is the product of this reaction. A synthesis reaction in a living organism is referred to as an *anabolic reaction* or *anabolism*. These are metabolic pathways.

- *Decomposition* reaction: A reaction that breaks the bond between atoms in a molecule or chemical compound. In $NaCl \rightarrow Na^+ + Cl^-$, sodium chloride is broken up into its chemical elements sodium and chlorine. A decomposition reaction in a living organism is called a *catabolic reaction* or *catabolism*.

- *Exchange* reaction: A reaction that is both a synthesis reaction and a decomposition reaction, where a chemical compound is decomposed into its chemical elements and those chemical elements are synthesized into a new chemical compound. In $NaOH + HCl \rightarrow NaCl + H_2O$, sodium hydroxide (NaOH) and hydrochloric acid (HCl) enter into an exchange reaction to form salt (NaCL) and water (H_2O).

A chemical reaction theoretically can be reversed, but in practice some reactions create an unstable chemical compound that might require special conditions to exist for the reverse reaction to happen. Those special conditions required to reverse a reaction appear below the arrow in the reaction notation. Above the arrow appears any special condition that must exist for the synthesized reaction to occur. In Fig. 2-6, a temperature of 250° C is the special condition for the synthesized reaction to occur and absolute zero is necessary for the decomposition reaction to occur.

A catalyst is a substance that speeds up the rate of a chemical reaction by decreasing the energy needed to run the reaction without changing the reactants or products. Enzymes are an example of a biological substance that acts as catalysts to speed up a reactor rate.

$$X + Y \underset{\text{water}}{\overset{\text{heat}}{\rightleftarrows}} XY$$

Fig. 2-6. In theory all chemical reactions are reversible. In practice these are called reversible reactions.

- *Velocity*. A specific level of energy is required for a bond to occur. This energy level is called *activation energy* and is different for each chemical reaction.
- *Orientation*. Two atoms, ions, or molecules must strike each other at a position where bonding can occur.
- *Reaction rate*. Collisions must occur frequently at the proper orientation and at the activation rate in order for bonding to happen. There are two ways to increase the reaction rate. These are an increase in temperature and an increase in pressure. Both cause atoms, ions, and molecules to move faster and increase the probability of a collision.
- *Size*. The atomic weight of an element influences the speed of a chemical reaction. An atom with a larger atomic weight than another atom requires more energy to be expended to increase the speed of the chemical reaction that binds the atom to another atom.

CATALYST: MAKING THINGS HAPPEN

Living organisms possess large molecules of proteins that are called *enzymes*. These enzymes act as *catalysts*. A catalysts is a chemical substance that speeds-up the rate of a chemical reaction. These catalysts do this without affecting the end products of the reaction, nor permanently altering themselves.

In order for an enzyme to be effective it must interact with a chemical called a *substrate*. The enzyme attaches itself to the substrate in an area that will most likely increase its ability to react. This *enzyme-substrate complex* lowers the activation energy of the reaction and enables the collision of chemicals involved in the reaction to be more effective.

An important factor of enzymes is that they can reduce the reaction time without the need to increase temperature. This is very important in living organisms because high temperatures can break apart the proteins that make up the cell.

CHEMICAL COMPOUND: MAKING SOMETHING USEFUL

As you learned previously in this chapter, a chemical element is a substance that cannot be divided into other chemical substances. For example, you cannot further divide hydrogen into anything because hydrogen is an element.

In its simplest form, a chemical element is made up of one atom. In its more complex form, a chemical element is made up of two or more atoms, which is called a molecule of the chemical element. For example, binding together two hydrogen atoms forms a hydrogen molecule.

$$H + H \rightarrow H_2$$

In order to make different things you need to combine different atoms and molecules of different chemical elements. This combination is called a compound. For example, combining two hydrogen molecules to an oxygen atom results in the compound we know as water.

$$H_2 + O \rightarrow H_2O$$

Molarity: Hey, There's a Mole Amongst Us

It seems nearly impossible to measure a molecule's mass or size. Fortunately, there is *Avagadro's number,* which is the number of particles in a mole of a substance. The number is 6.022×10^{23}. Amadeo Avagadro was an Italian physicist for whom the value was named.

Scientists can measure molecules using units called a mole. Abbreviated as *mol.* One *mole* is equal to the atomic weight of an element expressed in grams. A mole is the weight in grams of a substance that is equal to the sum of the atomic weights of the atoms in a molecule of the substance. This is referred to as a *gram molecular weight.*

Let's look at a water molecule to determine how many moles there are in a liter of water.

- Find the atomic mass for each chemical element that makes up water. Water has two chemical elements. These are hydrogen and oxygen.
- Look up the symbol for each element on the Periodic Table. These are H and O for hydrogen and oxygen.
- Note the bottom number alongside the symbol. This is the atomic mass for the chemical element. These are 1 for hydrogen and 16 for oxygen.
- Multiply the number of atoms of each chemical element in the molecule by its atomic mass to determine the value for one mole for the chemical element. For water, there are two hydrogen atoms so this will be $2 \times 1\,g$. One

mole of a hydrogen molecule H_2 equals 2 g. Water has one oxygen atom. Therefore, multiply 1×16 g. One mole of oxygen is 16 g.

- Sum the atomic mass of atoms that make up the molecule to determine one mole of the molecule. For water this is $2 g + 16 g = 18 g$. One mole of water equals the atomic mass of 18 g.

- The weight of one mole is the atomic mass of a molecule expressed in grams. Therefore, one mole of water weighs 18 grams.

- A liter of water has a mass of 1,000 grams. Calculate the number of moles per liter by dividing the number of grams (1,000) by one mole of water (18 grams). The result is 55 moles/liter.

An Unlikely Pair: Inorganic and Organic

Chemical compounds are divided into two general categories of substances. These are:

- *Inorganic compound.* A compound that does not contain the chemical element carbon (C).

- *Organic compound.* A compound containing carbon atoms, the exception is carbon dioxide (CO_2). Carbon dioxide is inorganic.

- Inorganic compounds are further divided into three categories. These are:

- *Acids.* An acid is any compound that dissociates into one or more hydrogen ions (H^+) and one or more negative ions (called anions) and is a proton donor.

- *Bases.* A base is any compound that dissociates into one or more positive ions (called cations) and one or more negative hydroxide ions. The negative hydroxide ions (OH^-) can either accept or share protons.

- *Salts.* A salt is an ionic compound that dissociates into one or more positive or negative ions in water, although some salts are not soluble in water. The positive and negative ions are neither hydrogen ions nor hydroxide ions. Sodium and chlorine atoms break away from the salt lattice when water molecules surround them. Water molecules become oriented so that the positive poles face the negatively charged chlorine ions and the negative poles face the positively charged sodium ions. The water's hydrogen shells react with the sodium and chlorine ions, drawing the ions from the salt lattice.

THE pH SCALE

There must be a balance between acids and bases in order to maintain life. An imbalance disrupts homeostasis. The acid-base balance is measured using the pH scale. The *pH scale* (Fig. 2-7) measures the acidity or alkalinity (base) of a substance using a pH value from 0 to 14. Values on the pH scale are logarithmic values. A pH value of 7 is neutral, which is the pH of pure water. A pH value greater than 7 is a base or alkaline. A pH value less than 7 is an acid. A change in one pH value is a large change because it is a logarithmic scale. For example a pH of 1 has 10 times more hydrogen ions than a pH of 2 and 100 times more hydrogen ions than a pH of 3 ($pH = -\log_{10}[H+]$).

Adding a substance that will increase or decrease the concentration of hydrogen ions can change the pH value of a substance. Increasing the concentration of hydrogen ions makes the substance more acidic and decreasing the concentration makes the substance more alkaline.

Fig. 2-7. The pH scale is a logarithmic scale that measures the acidity or alkalinity of a substance.

The pH value of chemical compounds in living things naturally fluctuates during metabolism. *Metabolism* is a collection of chemical reactions occurring in a living organism. Sometimes the chemical compound is more acidic than alkaline and vice versa. Any drastic sway in the acid-base balance could have a devastating effect. A chemical compound called a buffer is used to prevent harmful swings in the acid-base balance. A *buffer* releases hydrogen ions or binds hydrogen ions to stabilize the pH. A weak acid or base does not easily separate (ionization).

ORGANIC COMPOUNDS

An organic compound is a compound whose chemical elements include carbon. Carbon plays an import role in living things because compounds that contain it build many different organic compounds, each having different structures and functions. The large size of most carbon-containing molecules and the fact that they don't dissolve easily in water makes them useful in building body structures. Organic compounds also store energy required by an organism for metabolism.

Carbon can combine with other atoms because carbon has four electrons in its outer shell. This leaves room for four additional electrons from other atoms to bond to the carbon atom in a biological reaction (Fig. 2-8). Carbon also has low electronegativity and lacks polarity when a bond is formed.

A carbon atom commonly combines with other carbon atoms to form a *carbon chain*. There are two forms of carbon chains. These are *straight carbon chains* and *ring carbon chains*. Fig. 2-9 shows how this is used to illustrate fructose. Carbon chains are the basic form for many organic compounds.

Organic compounds come in many sizes—small to large. Many, but not all, large organic compounds are called *polymers*. A polymer is made up of small molecules called *monomers*. A monomer is another name for subunit. Monomers are bonded together to form a polymer in a process called *dehydration synthesis,* which removes water molecules from the compound.

Fig. 2-8.   The lines indicate a single bond with other atoms.

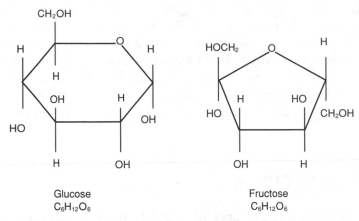

Fig. 2-9. A carbon cain is used to show the compound fructose.

A large organic compound is called a *macromolecule* which in many cases is a polymer. A *macromolecule* can be reduced to its monomer in a process called *hydrolysis*, which adds water molecules to the polymer.

There are four types of organic compounds that are macromolecules. These are:

- Carbohydrates
- Lipids
- Proteins
- Nucleic acids

Carbohydrates

Carbohydrates store energy from an organism in the form of sugar, starches and in the human body, glucogen. *Cellulose* is also a carbohydrate used as bulk to move food and waste through the gastrointestinal tract. Carbohydrates are also used as material in the cell wall. Carbohydrates are organized into three major Carbohydrate groups. These are:

- *Monosaccharides.* Some of the important monosaccharides are: glucose, the main energy source for an organism; fructose, acquired by eating fruit; galactose, which is in milk; deoxyribose, DNA; and ribose, RNA. A monomer is also a monosaccharide.

- *Disaccharides.* This is a combination of two monosaccharides bonded during dehydration synthesis. Sucrose (table sugar) and lactose (milk sugar) are disaccharides. Sucrose contains glucose and fructose. Lactose contains glucose and galactose.

- *Polysaccharides*. A polysaccharide is comprised of many monosaccharides and includes glycogen, starch, cellulose, and chitin, which is an amino sugar.

Lipids

Lipids are our fats and provide protection, insulation and can be used as an energy reserve. They are important components to the cell membrane and store pigments.

There are four kinds of lipids. These are:

- *Triglycerides*. Triglycerides protect and insulate the body from most lipids and are a source of energy. Because lipids have few polar covalent bonds, they are mostly insoluble (do not mix well) with polar solvents, like water.
- *Phospholipids*. Phospholipids are a major component in cell membranes.
- *Steroids*. Steroids are cholesterol and some hormones.
- *Eicosanoids*. Eicosanoids are divided into two kinds. These are prostaglandins and leukotrienes. Prostaglandins are involved in various behaviors such as dilating airways, regulating body temperature, and aiding in the formation of blood clots. Leukotrienes are involved in inflammatory and allergic responses.

Other kinds of lipids include fatty acids, lipoproteins and many plant pigments including chlorophyll and beta-carotene and the fat soluble vitamins such as A, D, E and K.

Proteins

Proteins comprise about 50% of a cell's dry weight and make up material in the cell wall. Proteins are peptidoglycans and help to transport chemicals into and out of a cell. In addition, proteins are part of cell structures and cytoplasmic components. Some proteins are antibodies that kill bacteria and play a role in muscle contractions and provide movement of microorganisms. Proteins are made up of polypeptides that bond together using peptide bonds. There are four structural levels of proteins. These are:

- *Primary*. The primary structure is the sequence in which amino acids are linked to form the polypeptide. Sequences are genetically determined and even the slightest alteration within the sequence may have a dramatic effect on the way the protein functions.

- *Secondary*. The secondary structure is locally folded and is the repeated twisting of the polypeptide chain that links together the amino acids. There are two types of secondary structures. These are a helix and a pleated sheet. The *alpha-helix* is a clockwise spiral structure. The *pleated sheet* forms the parallel portion of the polypeptide chain.

- *Tertiary*. The tertiary structure is the three-dimensional active structure of the polypeptide chain. Tertiary structure is the minimal level of structure for biological activity.

- *Quarternary*. Is where the proteins, in order to be functional, contain sub-units of polypeptide chains. An example would be DNA polymerase.

Proteins have many roles in a living organism. They are found in bone collagen and connective tissue and provide protection in the form of *immunoglobins,* which are antibodies. Some of the other important proteins are:

- *Myosin*. Muscle contraction.

- *Actin*. Muscle contraction.

- *Hemoglobin* transports oxygen (O_2) and carbon dioxide (CO_2) in blood.

- *Enzymes*. An enzyme that is a biological catalyst that increases the rate of chemical reactions in cells by reducing the energy required to begin the reaction. The reaction does not change the enzyme. The name of an enzyme typically ends with "-ase."

- *Flagellin*. Protein in flagella.

The Blueprint of Protein Synthesis

Proteins play a critical role in chemical reactions of microorganisms and other kinds of organisms. Information needed to direct the synthesis of a protein is contained in DNA (Deoxyribonucleic Acid). This information is transferred through generations from parent to child microorganisms. Nucleotides also store energy in high-energy bonds and form together to make nucleic acids.

There are three parts to a nucleotide:

- A nitrogen base, such as adenine.

- A five-carbon sugar, such as ribose.

- One or more phosphate groups.

Nucleic acids form by the joining of nucleotides that have stored energy that the microorganism needs for metabolism. Enzymes form to speed the rate of the chemical reaction that breaks these high-energy bonds to release the energy needed for cell metabolism.

Nucleic acids are long polymer chains that arc found in the nucleus of cells and contain all the genetic material of the cell. Genetic material determines the activities of the cell and is passed on from generation to generation.

TYPES OF NUCLEIC ACIDS

There arc two types of nucleic acids found in the cell:

- *Deoxyribonucleic acid (DNA)*. DNA is a double strand of nucleotides that is organized into segments. Each segment is called a gene. Genes determine the genetic markers that are inherited from previous generations of the organism. A genetic marker is a specific genetic characteristic such as

Table 2-3.   Scientists and Their Contributions

Year	Scientist	Contribution
1803	John Dalton	Developed the Atomic Theory that explains the relationship between an element and an atom.
1860	Dimitri Mendeleev	Organized chemical elements into a table according to its atomic weight.
1800	H. G. J. Moseley	Organized chemical elements into a table according to their atomic numbers.The table was originally know as the Law of Chemical Periodicity and has since been called the Periodic Table.
1911	Ernest Rutherford	Developed the Rutherford model of an atom and developed the concept of half-life.
1903	Marie and Pierre Curie	Discovered radioactivity.
1811	Amadeo Avagadro	An Italian physicist after whom the value of the number of molecules in a mole of a substance was named.
1913	Niels Bohr	Proposed that electrons occupy a cloud surrounding the nucleus of an atom. This is called an orbital.

the ability to synthesize proteins. Protein controls activities of the cell. Some microorganisms, such as viruses contain either DNA or RNA but not both. Think of DNA as a set of instructions.

- *Ribonucleic acid (RNA).* RNA is a single strand of nucleotides that relays instructions from genes to ribosomes, guiding the chemical reactions in the synthesis of amino acids into protein. Think of RNA as the person who carries out the instructions of DNA.

The Power House: ATP

Energy is stored in adenosine triphosphate (ATP) molecules. ATP supplies power necessary to:

- Move flagella in microorganisms.
- Move chromosomes in the cytoplasm.
- Transport substances in and out of the plasma membrane.
- Synthesis reactions.

ATP is synthesized from adenosine diphosphate (ADP) and a phosphate group (P), the latter of which gets its energy from the decomposition reaction of glucose and other substances. ATP releases energy in the form of ADP and P when ATP is decomposed.

Quiz

1. What is a dalton?
    (a) The equivalent of a nanometer
    (b) The unit of measurement used to measure the structure of an atom
    (c) The unit of measurement used to measure atomic number
    (d) The unit of measurement used to measure atomic weight

2. What is the name given to a chemical element whose atoms have a different number of neutrons?
    (a) Complex element
    (b) Differential element

(c) Isotope
(d) Stable element

3. How is the valence shell used?
    (a) The valence shell is used in bonding together two stable atoms.
    (b) The valence shell is used in bonding together two inorganic, stable atoms.
    (c) The valence shell is used in bonding together two organic, stable atoms.
    (d) The valence shell is used in bonding together two unstable atoms.

4. What process is used for two atoms to bond together?
    (a) Hydrogen synthesis
    (b) Ionic synthesis
    (c) A chemical reaction
    (d) Covalent synthesis

5. What is the difference between a molecule and a chemical compound?
    (a) A molecule consists of two or more atoms of different chemical elements. A chemical compound consists of two or more atoms of the same chemical element.
    (b) A molecule consists of only two atoms of the same chemical elements. A chemical compound consists of more than two atoms of different chemical elements.
    (c) A molecule consists of two or more atoms of the same chemical element. A chemical compound consists of two or more atoms of different chemical elements.
    (d) A molecule consists of two or more atoms of organic chemical element. A chemical compound consists of two or more atoms of inorganic chemical elements.

6. What kind of bond shares electrons in the valence shell?
    (a) Endergonic bond
    (b) Covalent bond
    (c) Hydrogen bond
    (d) Ionic bond

7. What kind of reaction uses chemical energy to bond atoms together?
    (a) Endergonic reaction
    (b) Fusion reaction
    (c) Fission reaction
    (d) Exchange reaction

8. What kind of reaction performs synthesis and decomposition?
    (a) Endergonic reaction
    (b) Fusion reaction
    (c) Fission reaction
    (d) Exchange reaction

9. What is the importance of orientation in a chemical reaction?
    (a) Atoms must be in the ideal position when they collide in order to have a high probability of bonding.
    (b) All atoms must be in a polar orientation in order for a chemical reaction to occur.
    (c) All atoms must be in a nonpolar orientation in order for a chemical reaction to occur.
    (d) Orientation is of no importance for a chemical reaction to occur.

10. How does an enzyme increase the likelihood that two atoms will bond?
    (a) An enzyme temporarily moves other atoms away, so it is highly likely that collision will occur.
    (b) An enzyme temporarily changes the chemical environment, so it is highly likely that collision will occur.
    (c) An enzyme temporarily bonds with an atom to help it move into a position where it is highly likely that it will collide and bond with another atom.
    (d) An enzyme temporarily changes the pH level, so it is highly likely that collision will occur.

Observing Microorganisms

Growing up you were probably forever being told to wash your hands so you would not become infected by germs. You probably complied only to stay out of trouble because no matter how well you focused on your hands, you never saw a germ on them. Today you realize that a germ is a microorganism, one of the many microorganisms that surround us. Microorganisms cannot be seen with the naked eye, but you can see them with the aide of a microscope. In this chapter, you'll learn how to view microorganisms under a microscope.

Size Is a Matter of Metrics

We could describe the size of a microorganism in a variety of ways. A microorganism is small, tiny, and miniscule. It is smaller than a human hair. Millions of them can fit on the head of a pin. All of these words give you an idea of how small a microorganism really is, but there's a problem using them to describe size. What is tiny? Are we talking about the diameter of a human hair or its length? As for the head of a pin, how big is the pin?

Words that are normally used to describe size do so in relative terms rather than provide a precise measurement. Relative terms compare one thing to another thing without scientific precision. For example the statement, "millions of them can fit on the head of a pin" isn't precise and raises a lot of questions. How many millions? What size is the head of the pin?

Speaking in relative terms is fine if you want to convey a general sense of size. Saying that millions of microorganisms can fit on the head of a pin gives someone a sense that a microorganism isn't the size of a dog or cat, but is something much smaller. However, scientists need to precisely measure the size of a microorganism in order to prevent diseases.

Let's see how this works by examining a surgical mask commonly used by medical professionals to control the flow of microorganisms. If you could zoom in on a surgical mask you would see that the surgical mask is a weave of threads that form a tiny web consisting of squareish holes. The size of each hole is determined by how close each strain of thread is to each other. Size is critical to reduce the spread of disease carrying microorganisms. If the hole is smaller than the microorganism, then the microorganism is unable to pass through the surgical mask. It becomes trapped or simply moves in a direction of less resistance. However a microorganism can easily pass through a hole that is larger than the microorganism.

Among other reasons, medical professionals choose surgical masks based on the size of microorganisms that they want to control. In order to make this selection, they need to precisely measure the size of a microorganism and the size of the holes created by the web of threads in the surgical mask.

NO FEET, PLEASE

Scientists measure the size of microorganisms—and practically everything else—by using metric measurements, commonly called the metric system. A *system* is a way of doing something, such as having a system to beat the odds in Las Vegas. The *metric system* is a way of measuring things by using multiples or fractions of ten, called *factors of ten* or *the power of ten*.

The metric system is the standard way of measuring things throughout the world except in the United States where we use the U.S. customary system of measurement, which includes inches, feet, and yards. The metric system is part of the the *Système International d'Unités* (SI system).

It is usually at this point in the study of microbiology when some students begin a slow panic because they must learn a new measurement system. Don't panic! The metric system is very easy to learn—much easier to learn than the U.S. measurement system.

THE PREFIX FIXES ALL YOUR PROBLEMS

The first trick to learning the metric system is to memorize the meaning of six prefixes. In the metric system, each prefix means that a meter is either multiplied or divided by a multiple of 10.

The second trick to learning the metric system is to learn how to multiply and divide by 10. This is easy because all you need to do is move the decimal point. The decimal point is moved to the right one place when multiplying by 10. The decimal point is moved to the left one place when dividing by 10.

Let's see how this works. First, multiply 1 meter by 10.

$$1 \times 10 = 10$$

Now convert a meter to a decimeter. From the previous section, you know that a decimeter is one-tenth of a meter, which means that you must divide by 10.

$$1 \div 10 = 0.10$$

Table 3-1 shows the prefixes for the metric system and their equivalent in meters.

Table 3-1.   Prefixes

Prefix	Value in Meters
Kilo (km) (kilo = 1,000)	1,000 m
Deci (dm) (deci = 1/10)	0.10 m
Centi (cm) (centi = 1/100)	0.01 m
Milli (mm) (milli = 1/1000)	0.001 m
Nano (nm) (nano = 1/1,000,000,000)	0.000000001 m
Pico (pm) (pico = 1/1,000,000,000,000)	0.000000000001 m

Table 3-2.   Units of Mass in the Metric System

Prefix	Value in Grams
Kilo (kg)	1,000 g
Hecto (hg)	100 g
Deka (dag)	10 g
Gram (g)	1 g
Deci (dg)	0.1 g
Centi (cg)	0.01 g
Milli (mg)	0.001 g
Micro (µg)	0.000001 g
Nano (ng)	0.000000001 g
Pico (pg)	0.000000000001 g

A meter is the standard for length in the metric system. A *kilogram* is the standard for mass in the metric system. A *gram* uses the same prefixes as a meter to specify the number of grams that are represented by a value. For example, a kilometer is 1,000 meters and a kilogram is 1,000 grams. This makes it a lot easier to learn the metric system since the number of grams and meters are indicated by the same set of prefixes.

Table 3-2 contains a list of various ways to express a gram. You'll notice that this table contains two prefixes that were not used in Table 3-1. These are deka- and hecto-. The prefix deka- means 10 and the prefix hecto- means 100. That's 10 grams and 100 grams.

SIZING UP MICROORGANISMS

How small is a microorganism? You are probably asking yourself this question after learning the metric system of measurement. The answer depends on the kind of microorganism that you are measuring. As you'll remember from Chapter 1, there are two general categories of microorganisms. These are prokaryotes and eukaryotes. Fig. 3-1 illustrates the relative size of a microorganism when compared to other things.

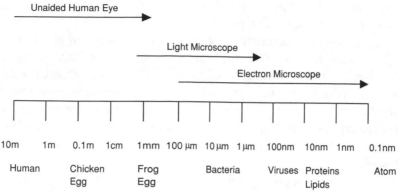

Fig. 3-1. Comparative sizes of humans and microorganisms.

Here's Looking at You

The principal way a microbiologist studies microorganisms is by observing them through a microscope. A *microscope* is a device that enlarges objects using a process called *magnification*. The simplest form of a microscope is a magnifying glass consisting of a single lens that is shaped in such a way as to make things appear larger than they are to the naked eye. And the simplest magnifying glass is the bottom of an empty glass. Some glasses are slightly bent at the bottom, causing a magnifying effect if held at a certain height over an object. This causes a change in the ray's path.

The most complex microscope is an electron microscope, which uses electrons to increase the apparent size of an object. Electron microscopes are capable of magnifying the organs of a microorganism called organelles, which you'll learn about later in this book. That is, an electron microscope is capable of showing what is inside a bacterium and virus.

WAVELENGTH

You don't really "see" anything. It sounds strange, but it is true. You see only the reflection of light waves—or, in the case of an electron microscope, the reflection or absence of electrons. Electromagnetic radiation is generated by a variety of sources, such as the sun, a light bulb, or a radio transmitter. It takes the form of a wave similar in shape to an ocean wave.

A wave has two characteristics. These are the wave height and the wavelength. The *wave height* is the highest level above the surface traveled by the wave. Let's say that you're traveling across a calm stretch of ocean. This is the surface. Your boat is then pushed high above the surface by a swell—the wave height—before returning to the surface. The *wavelength* is the distance between the highest point of two waves. That is, the distance the boat travels between the highest point of the first wave and the highest point of the second wave.

Waves of electromagnetic radiation are in a continuous scale and are clustered into groups called *bands*. They are given names based on their wavelengths. Some are probably familiar to you, such as x-rays, visible light waves, and radio waves. These groups are assembled into the electromagnetic spectrum.

Waves such as light waves are generated from a source such as the sun and strike an object such as your friend. Your friend absorbs some light waves and reflect other light waves. Your eyes detect only the reflected light waves.

It is this principle that enables you to observe a microorganism using a microscope. Light waves from either a light bulb or room light are reflected on to the microorganism. Reflected light waves are observed using the microscope. As you'll learn in Chapter 4, sometimes a microorganism reflects few light waves, making it difficult to see under a microscope. A *stain* is used to cause the microorganism to reflect different light waves. Microorganisms are visible under an electron microscope by directing waves of electrons onto the microorganism. Some electron waves are absorbed and others are reflected. The reflected waves are detected by an electronic circuit that displays an image of the microorganism on a video screen.

What Big Eyes You Have: Magnification

Light reflected from a specimen travels in a straight line to your eyes, which lets you see the specimen at its natural size. You can magnify the size of the specimen by looking at the specimen through a concave lens. A *convex lens* (Fig. 3-2) is usually made out of glass or plastic; the back of the lens is bent inward and the front is bent outward. It is like looking into a bubble. A magnifying glass is a convex lens.

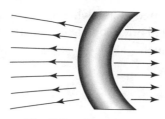

Fig. 3-2. Convex lens.

The specimen is the focal point, which is the place where all the reflected light originates. Light travels in a straight line from the focal point to the lens where the light is bent in a process called refraction. The angle at which light is bent is called the angle of refraction, which is measured in degrees from the natural path of the light. The degree of *angle of refraction* is determined by the curvature in the lens. The more the lens curves, the greater the angle of refraction.

The image appears larger as the light reflected from the image is refracted. Although the image appears magnified, curvature does distort the image. The amount of distortion depends on the angle of refraction and the distance between the lens and the specimen, which is called the *focal length*. The point at which light rays meet is called the *focal point* (Fig. 3-3).

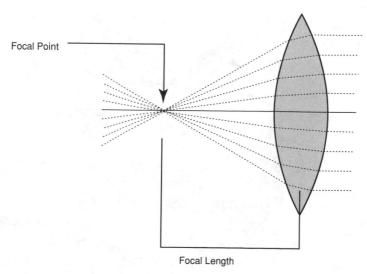

Fig. 3-3. The focal point is where light rays meet.

You probably saw a distorted image when using a magnifying glass and were able to minimize the distorted effect by changing the distance between the magnifying glass and the specimen.

The Microscope

A microscope is a complex magnifying glass. In the 1600s, during the time of Antoni van Leeuwenhoek (see Chapter 1), microscopes consisted of one lens that was shaped so that the refracted light magnified a specimen 100 times its natural size. Other lenses were shaped to increase the magnification to 300 times.

However, van Leeuwenhoek realized that a single-lens microscope is difficult to focus. Once Van Leeuwenhoek brought the specimen into focus, he kept his hands behind his back to avoid touching the microscope for fear they would bring the microscope out of focus. It was common in the 1600s for scientists to make a new microscope for each specimen that wanted to study rather than try to focus the microscope.

The single-lens magnifying lens or glass is a thing of the past. Scientists today use a microscope that has two sets of lenses (objective and ocular), which is called a compound light microscope. Fig. 3-4 shows parts of a compound light microscope. A compound light microscope consists of:

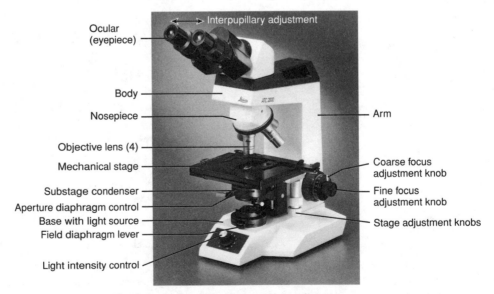

Fig. 3-4.   Parts of a compound light microscope.

- *Illuminator.* This is the light source located below the specimen.
- *Condenser.* Focuses the light through the specimen.
- *Stage.* The platform that holds the specimen.
- *Objective.* The lens that is directly above the stage.
- *Nosepiece.* The portion of the body that holds the objectives over the stage.
- *Field diaphragm.* Controls the amount of light into the condenser.
- *Base.* Bottom of the microscope.
- *Coarse focusing knob.* Used to make relatively wide focusing adjustments to the microscope.
- *Fine focusing knob.* Used to make relatively small adjustments to the microscope.
- *Body.* The microscope body.
- *Ocular eyepiece.* Lens on the top of the body tube. It has a magnification of 10× normal vision.

MEASURING MAGNIFICATION

A compound microscope has two sets of lenses and uses light as the source of illumination. The light source is called an *illuminator* and passes light through a *condenser* and through the specimen. Reflected light from the specimen is detected by the objective. The objective is designed to redirect the light waves, resulting in the magnification of the specimen.

There are typically four objectives, each having a different magnification. These are 4×, 10×, 40×, and 100×. The number indicates by how many times the original size of a specimen is magnified, so the 4× objective magnifies the specimen four times the specimen size. The eyepiece of the microscope is called the ocular eyepiece and it, too, has a lens—called an *ocular lens*—that has a magnification of 10×.

You determine the magnification used to observe a specimen under a microscope by multiplying the magnification of the objective by the magnification of the ocular lens. Suppose you use the 4× objective to view a specimen. The image you see through the ocular is 40× because the magnification of the object is multiplied by the magnification of the ocular lens, which is 10×.

Many microscopes have several objectives connected to a revolving nosepiece above the stage. You can change the objective by rotating the nosepiece until the objective that you want to use is in line with the body of the microscope. You'll find the magnification marked on the objective. Sometimes the mark is color-coded and other times the magnification is etched into the side of the objective.

RESOLUTION

The area that you see through the ocular eyepiece is called the *field of view*. Depending on the total magnification and the size of the specimen, sometimes the entire field of view is filled with the image of the specimen. Other times, only a portion of the field of view contains the image of the specimen.

You probably noticed that the specimen becomes blurry as you increase magnification. Here's what happens. The size of the field of view decreases as magnification increases, resulting in your seeing a smaller area of the specimen. However, the resolution of the image remains unchanged, therefore you must adjust the fine focus knob to bring the image into focus again.

Resolution is the ability of the lens to distinguish fine detail of the specimen and is determined by the wavelength of light from the illuminator. At the beginning of this chapter you learned about the wave cycle, which is the process of the wave going up and then falling down time and again. A wavelength is the distance between the peaks of two waves. As a general rule, shorter wavelengths produce higher resolutions of the image seen through the microscope.

CONTRAST

The image of a specimen must contrast with other objects in the field of view or with parts of the specimen itself to be visible in different degrees of brightness. Suppose the specimen was a thin tissue layer of epidermis. The tissue must be a different color than the field of view, otherwise the tissue and field of view blend, making it impossible to differentiate between the two. That is, the tissue and the field of view must contrast.

Previously in this chapter you learned that what you see is light reflected by the specimen (or the transmitted light if the specimen doesn't absorb light). The illuminator shines white light onto the specimen. White light contains all the light waves in the visible spectrum. The specimen absorbs some of the light waves and reflects other light waves, giving the appearance of some color other than white.

Light waves that are reflected by the specimen are measured by the refractive index. The *refractive index* specifies the amount of light waves that is reflected by an object. There is a low contrast between a specimen and the field of view if they have nearly the same refractive index. The further these refractive indexes are from each other, the greater the contrast between the specimen and the field of view.

Unfortunately, refractive indexes of the specimen and the field of view are fixed. However, you can tweak the refractive index of the specimen by using a stain. The stain adheres to all or part of the specimen, absorbing additional light waves and increasing the difference between the refractive indexes of the specimen and the field of view. This results in an increase in the contrast between the specimen and the field of view.

Oil Immersion

A challenge facing microbiologists is how to maintain good resolution at magnifications of 100× and greater. In order to maintain good resolution, the lens must be small and sufficient light must be reflected from both the specimen and the stain used on the specimen. The problem is that too much light is lost; air between the slide and the objective prevents some light waves from passing to the objective, causing the fuzzy appearance of the specimen in the ocular eyepiece.

The solution is to immerse the specimen in oil. The oil takes the place of air and, since oil has the same refractive index as glass, the oil becomes part of the optics of the microscope. Light that is usually lost because of the air is no longer lost. The result is good resolution under high magnification.

TYPES OF LIGHT COMPOUND MICROSCOPES

There are five popular light compound microscopes used today (see Table 3-3).

Bright-Field Microscope

The bright-field microscope is the most commonly used microscope and consists of two lenses. These are the ocular eyepiece and the objective. Light coming from the illuminator passes through the specimen. The specimen absorbs some light waves and passes along other light waves into the lens of the microscope, causing a contrast between the specimen and other objects in the field of view. Specimens that have pigments contrast with objects in the field of view and can be seen by using the bright-field microscope. Specimens with few or no pigments have a low contrast and cannot be seen with the bright-field microscope. Some bacteria have low contrast.

Table 3-3.   Quick Guide to Microscopy

Type of Microscope	Features	Best Used for
Bright-field	Uses visible light	Observing dead stained specimens and living organisms with natural color
Dark-field	Uses visible light with a that causes the rays of light to reflect off the specimen	Observing living organisms
Phase-contrast	Uses a condenser that increases differences in the refractive index of structures within the specimen	Observing internal structures of specimens
Fluorescent	Uses ultraviolet light to stimulate molecules of the specimen to make it stand out from its background	Observing specimens or antibodies in clinical studies
Transmission electron microscope	Uses electron beams and electromagnetic lenses to view thin slices of cells	Observing exterior surfaces and internal structures
Scanning electron microscope	Uses electron beams and electromagnetic lenses	Giving a three-dimensional view of exterior surfaces of cells

Dark Field Microscope

The dark-field microscope focuses the light from the illuminator onto the top of the specimen rather than from behind the specimen. The specimen absorbs some light waves and reflects other light waves into the lens of the microscope. The field of view remains dark while the specimen is illuminated, providing a stark contrast between the field of view and the specimen.

Phase-Contrast Microscope

The phase-contrast microscope bends light that passes through the specimen so that it contrasts with the surrounding medium. Bending the light is called moving the light out of phase. Since the phase-contrast microscope compensates for the refractive properties of the specimen, you don't need to stain the specimen to enhance the contrast of the specimen with the field of view. This microscope is ideal for observing living microorganisms that are prepared in wet mounted slides so you can study a living microorganism.

Fluorescent Microscope

Fluorescent microscopy uses ultraviolet light to illuminate specimens. Some organism fluoresce naturally, that is, give off light of a certain color when exposed to the light of different color. Organisms that don't fluoresce naturally can be stained with fluorochrome dyes. When these organisms are placed under a fluorescent microscope with an ultraviolet light, they appear very bright in front of a dark background.

Differential Interface Contrast Microscope (Nomanski)

The differential interface contrast microscope, commonly known as Nomanski, works in a similar way to the phase-contrast microscope. However, unlike the phase-contrast microscope (which produces a two-dimensional image of the specimen), the differential interface contrast microscope shows the specimen in three dimensions.

THE ELECTRON MICROSCOPE

A light compound microscope is a good tool for observing many kinds of microorganisms. However, it isn't capable of seeing the internal structure of a microorganism nor can it be used to observe a virus. These are too small to effectively reflect visible light sufficient to be seen under a light compound microscope. In order to view internal structures of viruses and internal structures of microorganisms, microbiologists use an electron microscope where specimens are viewed in a vacuum.

Developed in the 1930s, the electron microscope uses beams of electrons and magnetic lenses rather than light waves and optical lenses to view a specimen. Very thin slices of the specimen are cut so that the internal structures can be viewed. Microscopic photographs called *micrographs* are taken of the specimen and viewed on a video screen. Specimens can be viewed up to 200,000 times normal vision. However, living specimens cannot be viewed because the specimen must be sliced.

Transmission Electron Microscope

The transmission electron microscope (TEM) has a total magnification of up to $200,000\times$ and a resolution as fine as seven nanometers. A nanometer is $1/1,000,000,000$ of a meter. The transmission electron microscope generates an image of the specimen two ways. First, the image is displayed on a screen similar to that of a computer monitor. The image can also be displayed in the form

of an electron micrograph, which is similar to a photograph. Specimens viewed by the transmission electron microscope must be cut into very thin slices, otherwise the microscope does not adequately depict the image.

Scanning Electron Microscope

The scanning electron microscope (SEM) is less refined than the transmission electron microscope. It can provide total magnification up to 10,000× and a resolution as close as 20 nanometers. However, a scanning electron microscope produces three-dimensional images of specimen. The specimen must be freeze-dried and coated with a thin layer of gold, palladium, or other heavy metal.

Preparing Specimens

There are two ways to prepare a specimen to be observed under a light compound microscope. These are a smear and a wet mount.

Smear

A *smear* is a preparation process where a specimen that is spread on a slide. You prepare a smear using the *heat fixation process:*

1. Use a clean glass slide.
2. Take a loop of the culture.
3. Place the live microorganism on the glass slide.
4. The slice is air dried then passed over a Bunsen burner about three times.
5. The heat causes the microorganism to adhere to the glass slide. This is known as fixing the microorganism to the glass slide.
6. Stain the microorganism with an appropriate stain (see "Staining a Specimen" later in this chapter).

Wet Mount

A wet mount is a preparation process where a live specimen in culture fluid is placed on a concave glass side or a plain glass slide. The concave portion of the glass slide forms a cup-like shape that is filled with a thick, syrupy substance, such as *carboxymethyl cellulose*. The microorganism is free to move about within the fluid, although the viscosity of the substance slows its movement. This makes it easier for you to observe the microorganism. The specimen and the substance are protected from spillage and outside contaminates by a glass cover that is placed over the concave portion of the slide.

STAINING A SPECIMEN

Not all specimens can be clearly seen under a microscope. Sometimes the specimen blends with other objects in the background because they absorb and reflect approximately the same light waves. You can enhance the appearance of a specimen by using a stain. A stain is used to contrast the specimen from the background.

A *stain* is a chemical that adheres to structures of the microorganism and in effect dyes the microorganism so the microorganism can be easily seen under a microscope. Stains used in microbiology are either basic or acidic.

Basic stains are cationic and have positive charge. Common basic stains are methylene blue, crystal violet, safranin, and malachite green. These are ideal for staining chromosomes and the cell membranes of many bacteria.

Acid stains are anionic and have a negative charge. Common acidic stains are eosin and picric acid. Acidic stains are used to stain cytoplasmic material and organelles or inclusions.

Types of Stains

There are two types of Stains: simple and differential. See Table 3-4 for a summary of staining techniques.

Simple Stain

A simply stain has a single basic dye that is used to show shapes of cells and the structures within a cell. Methylene blue, safranin, carbolfuchsin and crystal violet are common simple stains that are found in most microbiology laboratories.

Differential Stain

A differential stain consists of two or more dyes and is used in the procedure to identify bacteria. Two of the most commonly used differential stains are the Gram stain and the Ziehl-Nielsen acid-fast stain.

In 1884 Hans Christian Gram, a Danish physician, developed the Gram stain. Gram-stain is a method for the differential staining of bacteria. Gram-positive microorganisms stain purple. Gram-negative microorganisms stain pink. *Staphylococcus aureus*, a common bacterium that causes food poisoning, is gram-positive. *Escherichia coli* is gram-negative.

Table 3-4. Quick Guide for Staining Techniques

Type	Number of Dyes Used	Observations	Examples
Simple stains	Uses a single dye	Size, shape, and arrangement of cells	Methylene blue Safranin Crystal violet
Differential stains	Uses two or more dyes to distinguish different types or different structures of bacteria	Distinguish gram-positive or gram-negative Distinguishes the members of mycobacteria and nocardia from other bacteria	Gram stain Ziehl-Nielsen acid-fast stain
Special stains	These stains identify specialized structures	Exhibit the presence of flagellae Exhibits endospores	Shaeffer-Fulton spore staining

The *Ziehl-Nielsen acid-fast stain*, developed by Franz Ziehl and Friedrick Nielsen, is a red dye that attaches to the waxy material in the cell walls of bacteria such as *Mycobacterium tuberculosis*, which is the bacterium that causes tuberculosis, and *Mycobacterium leprae*, which is the bacterium that causes leprosy. Microorganisms that retain the Ziehl-Nielsen acid-fast stain are called *acid-fast*. Those that do not retain it turn blue because the microorganism doesn't absorb the Ziehl-Nielsen acid-fast stain.

Here's how to Gram-stain a specimen (Fig. 3-5).

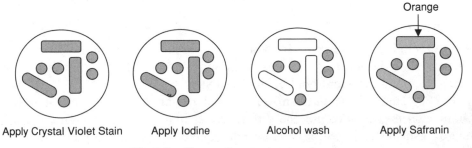

Orange

Apply Crystal Violet Stain   Apply Iodine   Alcohol wash   Apply Safranin

Fig. 3-5. How to Gram-stain a specimen.

1. Prepare the specimen using the heat fixation process (see "Smear" earlier in this chapter).

2. Place a drop of crystal violet stain on the specimen.

3. Apply iodine on the specimen using an eyedropper. The iodine helps the crystal violet stain adhere to the specimen. Iodine is a mordant, which is a chemical that fixes the stain to the specimen.

4. Wash the specimen with ethanol or an alcohol-acetone solution, then wash with water.

5. Wash the specimen to remove excess iodine. The specimen appears purple in color.

6. Wash the specimen with an ethanol or alcohol-acetone decolorizing solution.

7. Wash the specimen with water to remove the dye.

8. Apply the safranin stain to the specimen using an eyedropper.

9. Wash the specimen.

10. Use a paper towel and blot the specimen until the specimen is dry.

11. The specimen is ready to be viewed under the microscope. Gram-positive bacteria appear purple and gram-negative bacteria appear pink.

Here's how to apply the Ziehl-Nielsen acid-fast stain to a specimen.

1. Prepared the specimen (see "Smear" earlier in this chapter).

2. Apply the red dye carbol-fuchsin stain generously using an eyedropper.

3. Let the specimen sit for a few minutes.

4. Warm the specimen over steaming water. The heat will cause the stain to penetrate the cell wall.

5. Wash the specimen with an alcohol-acetone decolorizing solution consisting of 3 percent hydrochloric acid and 95 percent ethanol. The hydrochloric acid will remove the color from non–acid-fast cells and the background. Acid-fast cells will stay red because the acid cannot penetrate the cell wall.

6. Apply methylene blue stain on the specimen using an eyedropper.

Special Stains

Special stains are paired to dye specific structures of microorganisms such as endospores, flagella, and gelatinous capsules. One stain in the pair is used as a negative stain. A *negative stain* is used to stain the background of the micro-

Table 3-5.   Scientists and Their Contributions

Year	Scientists	Contribution
1884	Hans Christian Gram	Developed the Gram stain used to stain and identify bacteria.
	Franz Ziehl and Friedrick Nielsen	Developed the Ziehl-Nielsen acid-fast stain used to stain bacteria.

organism, causing the microorganism to appear clear. A second stain is used to colorize specific structures within the microorganism. For example, nigrosin and India ink are used as a negative stain and methylene blue is used as a positive stain.

The *Schaeffer-Fulton endospore stain* is a special flagellar stain that is used to colorize the endospore. The *endospore* is a dormant part of the bacteria cell that protects the bacteria from the environment outside the cell.

Here's how to apply the Schaeffer-Fulton endospore stain.

1. Prepare the specimen (see "Smear" earlier in this chapter).
2. Heat the malachite green stain over a Bunsen burner until it becomes fluid.
3. Apply the malachite green to the specimen using an eyedropper.
4. Wash the specimen for 30 seconds.
5. Apply the safranin stain using an eyedropper to the specimen to stain parts of the cell other than the endospore.
6. Observe the specimen under the microscope.

Quiz

1. What is a nanometer?
   (a) 1/1,000,000,000 of a meter
   (b) 1/100,000 of a meter
   (c) 1/1,000,000 of a meter
   (d) 1,000,000,000 meters

2. What magnification is used if you observe a microorganism with a microscope whose object is 100× and whose ocular lens is 10×?
   (a) 1000× magnification
   (b) 100× magnification
   (c) 10× magnification
   (d) 10,000× magnification

3. What is the function of an illuminator?
   (a) To control the temperature of the specimen
   (b) To keep the specimen moist
   (c) An illuminator is the light source used to observe a specimen under a microscope
   (d) To keep the specimen dry

4. What is the area seen through the ocular eyepiece called?
   (a) The stage
   (b) The objective
   (c) The display
   (d) The field of view

5. How do you maintain good resolution of a specimen at magnifications greater than 100×?
   (a) Display the specimen on a television monitor.
   (b) Use a single ocular eyepiece.
   (c) Immerse the specimen in oil.
   (d) Avoid moving the specimen.

6. What is a micrograph?
   (a) A microscopic photograph taken by an electron microscope
   (b) A microscopic diagram of a specimen
   (c) A microscopic photograph taken by a light microscope
   (d) A growth diagram of a specimen

7. What is a smear?
   (a) A smear is a preparation process in which a specimen is spread on a slide.
   (b) A smear is a preparation process in which a specimen is dyed.
   (c) A smear is a process in which a specimen is moved beneath a microscope.
   (d) A smear is a process used to identify a specimen.

8. What process is used to cause a specimen to adhere to a glass slide?
   (a) The heat fixation process
   (b) Wet mount
   (c) White glue
   (d) Clear glue

9. Why is a specimen stained?
   (a) A stain is used to label a specimen.
   (b) A stain is used to determine the size of a specimen.
   (c) A stain adheres to the specimen, causing more light to be reflected by the specimen into the microscope.
   (d) A stain is used to determine the density of a specimen.

10. When would you use a wet mount?
   (a) A wet mount is used to observe a dead specimen under a microscope.
   (b) A wet mount is used to observe a live specimen under a microscope.
   (c) A wet mount is used to observe an inorganic specimen under a microscope.
   (d) A wet mount is the first step in preparing a specimen.

CHAPTER

4

Prokaryotic Cells and Eukaryotic Cells

What do you and *Athletes Foot* have in common? You'll recall from Chapter 1 that *Tinea pedis* is the scientific name for athlete's foot and that it is caused by the *Trichophyton* fungus. Both of you are alive. Sometimes it's hard to imagine that microorganisms are alive because we can't see them with our naked eye—although they make their presence known to us in annoying ways.

Humans, *Trichophyton*, and all other living things are alive because they carry on the six life processes, something non-living things do not do. The six life processes require a living thing to:

- *Metabolize.* Breakdown nutrients for energy or extract energy from the environment.
- *Be responsive.* React to internal and external environmental changes.
- *Move.* Whether it is the entire organism relocating within its environment, cells within that organism or the organelles inside those cells.
- *Grow.* Increase the size or number of cells.
- *Differentiate.* The process where cells that are unspecialized become specialized. (An example would be a single fertilized human egg, developing into an individual). Prokaryotic cells do not differentiate.
- *Reproduce.* Form new cells to create a new individual.

Table 4-1.   Basic Life Processes in Microorganisms

Process	Eukaryotic Cells	Prokaryotic Cells	Viruses
Metabolism: Sum of all chemical reactions	Yes	Yes	Uses their host's cells for metabolism
Responsiveness: Ability to react to environmental stimuli	Yes	Yes	Some viruses react and multiply once they enter a host cells
Movement: Motion of individual organelles, a single cell, or entire organism	Yes	Yes	Virions (viruses outside of a host cell) are nonmotile. Use Brownian movement (random collision)
Growth: Increase in size	Yes	Yes	No
Differentiate	Yes	No	No
Reproduction: Increase in number	Yes	Yes	Inside host cells
Cellular structures	Yes	Yes	No cytoplasmic membrane No cell structure

Have a cellular structure. Cells that metabolize, respond to changes, move, grow, and reproduce.

For additional information, see Table 4-1.

In this chapter, you will learn about cellular structure by exploring two kinds of cells. These are prokaryotic cells and eukaryotic cells. Bacteria are prokaryotic organisms. Animals, plants, algae, fungi and protozoa are eukaryotic organisms. See Table 4-2 for a summary of differences between prokaryotic and eukaryotic cells.

Prokaryotic Cells

A *prokaryotic cell* is a cell that does not have a true nucleus. The nuclear structure is called a nucleoid. The *nucleoid* contains most of the cell's genetic mate-

Table 4-2. Differences between Prokaryotic and Eukaryotic Cells

Characteristics	Prokaryotic Cells	Eukaryotic Cells
Cell wall	Include peptidoglycan Chemically complex	Chemically simple
Plasma membrane	No carbohydrates No sterols	Contain carbohydrates Contain sterols
Glycocalyx	Contain a capsule or a slime layer	Contained in cells that lack a cell wall
Flagella	Protein building blocks	Multiple microtubules
Cytoplasm	No cytoplasmic streaming	Contain cytoskeleton Contain cytoplasmic streaming
Membrane-bound organelles	None	Endoplasmic reticulum Golgi complex Lysomes Mitochondria Chloroplasts
Ribosomes	70S	80S Ribosomes located in Organelles are 70S
Nucleus	No nuclear membrane No nucleoli 0.2–2.0 mm in diameter	Have a nucleus Have a nuclear membrane Have a nucleoli 10–100 mm in diameter
Chromosomes	Single circular chromosome No histones	Multiple linear chromosomes Have histones
Cell division	Binary fission	Mitosis
Sexual reproductions	No meiosis DNA transferred in fragments	Meiosis

rial and is usually a single circular molecule of DNA. *Karyo-* is Greek for "kernel." A prokaryotic organism, such as a bacterium, is a cell that lacks a membrane-bound nucleus or membrane-bound organelles. The exterior of the cell usually has glycocalyx, flagellum, fimbriae, and pili (Fig. 4-1).

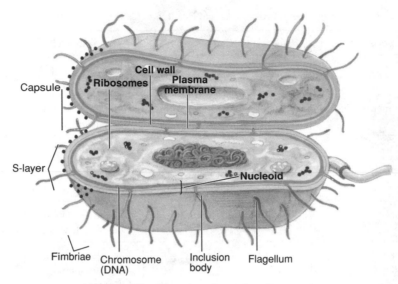

Fig. 4-1. A prokaryotic cell.

PARTS OF PROKARYOTIC CELLS

Glycocalyx

Glycocalyx is a sticky, sugary envelope composed of polysaccharides and/or polypeptides that surround the cell. Glycocalyx is found in one of two states. It can be firmly attached to the cell's surface, called *capsule*, or loosely attached, called *slime layer*. A slime layer is water-soluble and is used by the prokaryotic cell to adhere to surfaces external to the cell.

Glycocalyx is used by a prokaryotic cell to protect it against attack from the body's immune system. This is the case with *Streptococcus mutans*, which is a bacterium that colonizes teeth and excretes acid that causes tooth decay. Normally the body's immune system would surround the bacterium and eventually kill it, but that doesn't happen with *Streptococcus mutans*. It has a glycocalyx capsule state, which prevents the *Streptococcus mutans* from being recognized as a foreign microorganism by the body's immune system. This results in cavities.

Flagella

Flagella (Fig. 4-2) are made of protein and appear "whip-like." They are used by the prokaryotic cell for mobility. Flagella propel the microorganism away from

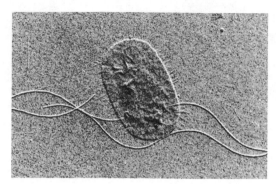

Fig. 4-2.   A flagellum has a long tail that extends from the cell.

harm and towards food in a movement known as *taxis*. Movement also occurs in response to a light or chemical stimulus. Movement caused by a light stimulus is referred to as *phototaxis* and a chemical stimulus causes a *chemotaxis* movement to occur.

Flagella can exist in the following forms:

- Monotrichous: One flagellum.
- Lophotrichus: A clump of flagella, called a tuft, at one end of the cell.
- Amphitrichous: Flagella at two ends of the cell.
- Peritrichous: Flagella covering the entire cell.
- Endoflagellum: A type of amphitrichous flagellum that is tightly wrapped around spirochetes. A *spirochete* is a spiral-shaped bacterium that moves in a corkscrew motion. *Borrelia burgdorferi*, which is the bacterium that causes lyme disease, exhibits an endoflagellum.

Fimbriae

Fimbriae are proteinaceous, sticky, bristle-like projections used by cells to attach to each other and to objects around them. *Neisseria gonorrhoeae*, the bacterium that causes gonorrhea, uses fimbriae to adhere to the body and to cluster cells of the bacteria.

Pili

Pili are tubules that are used to transfer DNA from one cell to another cell similar to tubes used to fuel aircraft in flight. Some are also used to attach one cell to another cell. The tubules are made of protein and are shorter in length than flagella and longer than fimbriae.

CELL WALL

The prokaryotic cell's cell wall is located outside the plasma membrane and gives the cell its shape and provides rigid structural support for the cell. The cell wall also protects the cell from its environment.

Pressure within the cell builds as fluid containing nutrients enters the cell. It is the job of the cell wall to resist this pressure the same way that the walls of a balloon resist the build-up of pressure when it is inflated. If pressure inside the cell becomes too great, the cell wall bursts, which is referred to as lysis.

The cell wall of many bacteria is composed of peptidoglycan, which covers the entire surface of the cell. *Peptidoglycan* is made up of a combination of peptide bonds and carbohydrates, either N-acetylmuramic acid, commonly referred to as *NAM,* or N-acetylglucosamine, which is known as *NAG.*

The wall of a bacterium is classified in two ways:

- *Gram-positive.* A gram-positive cell wall (Fig. 4-3) has many layers of peptidoglygan that retain the crystal of violet dye when the cell is stained. This gives the cell a purple color when seen under a microscope.

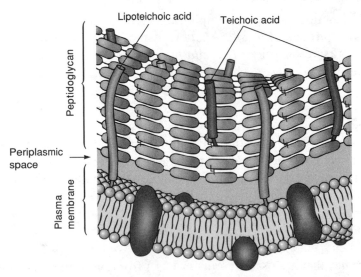

Fig. 4-3. Gram-positive cell wall.

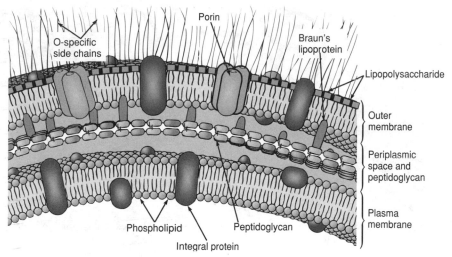

Fig. 4-4.   Gram-negative cell wall.

- *Gram-negative.* A gram-negative cell wall (Fig. 4-4) is thin. The inside is made of peptidoglycan. The outer membrane is composed of phospholipids and lipopolysaccharides.

The cell wall does not retain the crystal of violet dye when the cell is stained. The cell appears pink when viewed with a microscope.

CYTOPLASMIC MEMBRANE

The prokaryotic cell has a cell membrane called the *cytoplasmic membrane* that forms the outer structure of the cell and separates the cell's internal structure from the environment. The cytoplasmic membrane is a membrane that provides a selective barrier between the environment and the cell's internal structures.

The cytoplasmic membrane (Fig. 4-5) provides a selective barrier, allowing certain substances and chemicals to move into and out of the cell. The cytoplasmic membrane is a bilayer of phospholipids that has polar and nonpolar parts, which is referred to as being *amphipathic*. The nonpolar parts share electrons of atoms equally. The polar parts share electrons unequally. Each polar part has a head that contains phosphate and is hydrophilic ("water-loving"). Each nonpolar part has two tails composed of long fatty acids that are hydrophobic ("water-fearing").

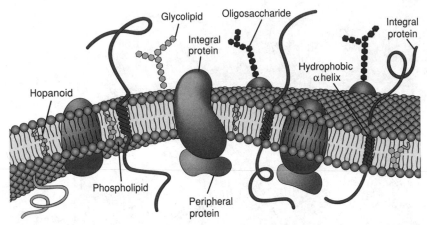

Fig. 4-5.   The cytoplasmic membrane enables some substances to pass into and out of the cell.

The heads always face a watery fluid such as the extracellular fluid on the outside of the cell and the intracellular fluid inside the cell. Tails align back to back preventing the watery fluid from crossing the cytoplasmic membrane.

The Fluid Mosaic Model

In 1972, S. J. Singer and G. L. Nicolson developed the *Fluid Mosaic Model,* which describes the structure of the cytoplasmic membrane. They called this a mosaic because proteins within the cytoplasmic membrane are arranged like tiles in a mosaic artwork. The term fluid is used to imply that membrane proteins and lipids flow freely within the cytoplasmic membrane. There are two kinds of proteins within the cytoplasmic membrane. These are:

- *Integral proteins.* An integral protein extends into the lipid bilayer. Integral proteins are typically glycoproteins that act like a molecular signature that cells use to recognize each other. Glycoproteins have a carbohydrate group attached to them. Two examples are:

  - *Transmembrane protein.* A transport protein that regulates the movement of molecules through the cytoplasmic membrane.

  - *Channel protein.* A channel protein forms pores or channels in the cytoplasmic membrane that permit the flow of molecules through the cytoplasmic membrane.

- *Peripheral proteins.* Peripheral proteins are on the inner and outer surface of the cytoplasmic membrane and have the characteristics of a polar and non-polar regions.

The Function of the Cytoplasmic Membrane

The cytoplasmic membrane regulates the flow of molecules (such as nutrients) into the cell and removes waste from the cell by opening and closing passages called channels. In photosynthetic prokaryotes, the cytoplasmic membrane functions in energy production by collecting energy in the form of light.

The cytoplasmic membrane is *selectively permeable* because it permits the transport of some substances and inhibits the transport of other substances. Two types of transport mechanisms are used to move substances through the cytoplasmic membrane. These are passive transport and active transport.

Passive Transport

Passive transport moves substances into and out of the cell down a gradient similar to how a rock rolls downhill, following the gradient. There are three types of passive transport. These are:

- *Simple diffusion.* Simple diffusion is the movement of substances from a higher-concentration region to a lower-concentration region (net movement). Only small chemicals (oxygen and carbon dioxide) or lipid-soluble chemicals (fatty acids) diffuse freely through the cytoplasmic membrane, using simple diffusion. Large molecules (monosaccharide and glucose) are too large to enter the cell.

- *Facilitated diffusion.* Facilitated diffusion is the movement of substances from a higher-concentration region to a lower-concentration region (net movement) with the assistance of an integral protein across a *selectively permeable membrane*. The phospholipid bilayer prevents the movement of large molecules across the membrane until a pathway is formed using facilitated diffusion. The integral protein acts as a carrier by changing the shape of large molecules so the protein can transport the large molecules across the membrane.

- *Osmosis.* Osmosis is the net movement (diffusion) of a solvent (water in living organisms) from a region of higher water concentration to a region of lower concentration.

How Osmosis Works

- *Isotonic solution.* "Iso" means the same if a cell is placed in an isotonic solution. This means that there is the same concentration of solute and solvent (water) inside and outside of the cell. There is an *equal* movement of substances into and out of the cell.
- *Hypertonic solution.* In a hypertonic solution, the cell is placed in an environment where there is a higher concentration of solute. What happens is that the water inside the cell will move out of the cell by osmosis causing it to shrink. This shrinking of the cell is called crenation.
- *Hypotonic solution.* When a cell is placed in a hypotonic solution, there is more water outside of the cell than inside. This means there is more solute concentration inside of the cell. The water outside will move into the cell by osmosis causing the cell to swell and ultimately break apart. This is called lysis.

Active Transport

Active transport is the movement of a substance across the cytoplasmic membrane against the gradient by using energy provided by the cell. This is similar to pumping water against gravity through a pipe. Energy must be spent in order for the pump to work.

A cell makes energy available by removing a phosphate (P) from *adenosine triphosphate* (ATP). ATP contains chemical potential energy that is released by a chemical reaction within the cell. It is this energy that is used to change the shape of the integral membrane protein-enabling substances inside the cell to be pumped through the cytoplasmic membrane. For example, active transport is used to pump sodium (Na^+) from a cell.

Group Translocation

Group translocation is a diffusion process that immediately modifies a substance once the substance passes through the cytoplasmic membrane. The cell must expend energy during group translocation, which is supplied by high-energy phosphate compounds such as phosphoenolpyruvic acid (PEP). Group translocation occurs in prokaryotic cells.

Endocytosis and Exocytosis

Endocytosis and exocytosis are processes used to move large substances or lots of little ones into and out of a cell. Large substances enter the cell by *endocyto-*

sis. There are two kinds of endocytosis. These are phagocytosis and pinocytosis. *Phagocytosis* engulfs solid substances (large molecules) while *pinocytosis* engulfs liquid substances (small molecules). *Exocytosis* is the process that cells use to remove large substances, which is the way waste products and useful material as hormones and neurotransmitters are expelled from a cell through *vesicles*.

Cytosol and Cytoplasm

The *cytosol* is the intracellular fluid of a prokaryotic cell that contains proteins, lipids, enzymes, ions, waste, and small molecules dissolved in water, commonly referred to as *semifluid*. Substances dissolved in cytosol are involved in cell metabolism.

The cytosol also contains a region called the *nucleoid*, which is where the DNA of the cell is located. Unlike human cells, a prokaryotic microorganism has a single chromosome that isn't contained within a nuclear membrane or envelope.

Cytosol is located in the cytoplasm of the cell. *Cytoplasm* also contains the cytoskeleton, ribosomes, and inclusions.

Ribosomes

A *ribosome* is an organelle within the cell that synthesizes polypeptide. There are thousands of ribosomes in the cell. You'll notice them as the grainy appearance of the cell when viewing the cell with an electron microscope.

A ribosome is comprised of subunits consisting of protein and *ribosomal RNA*, which is referred to as *rRNA*. Ribosomes and their subunits are identified by their sedimentation rate. *Sedimentation rate* is the rate at which ribosomes are drawn to the bottom of a test tube when spun in a centrifuge. Sedimentation rate is expressed in *Svedberg (S) units*. A sedimentation rate reflects the mass, size, and shape of a ribosome and its subunits. It is for this reason why the sedimentation rates of subunits of a ribosome do not add up to the ribosome's sedimentation rate.

Ribosomes in prokaryotic cells are uniquely identified by the number of proteins and rRNA molecules contained in the ribosome and by sedimentation rate. Prokaryotic ribosomes are relatively small and less dense than ribosomes of other microorganisms. For example, bacterial ribosomes have a sedimentation rate of 70S compared to the 80S sedimentation rate of a eukaryotic ribosome, which you'll learn about later in this chapter.

Ribosomes and their subunits are targets for antibiotics that kill a bacterium by inhibiting the bacterium's protein synthesis. These antibiotics only kill cells that have a specific ribosome sedimentation rate. Cells with a different ribosome sedimentation rate are unaffected by the antibiotic. This enables the antibiotic to kill bacterium and not the body that is infected by the bacterium.

For example, erythromycin and chloramphenicol, popular antibiotics, kill bacteria whose subunits have a sedimentation rate of 50S. Streptomycin and gentamycin affect bacteria whose subunits have a 30S sedimentation rate.

Inclusions

An inclusion is a storage area that serves as a reserve for lipids, nitrogen, phosphate, starch, and sulfur within the cytoplasm. Scientists use inclusions to identify types of bacteria. Inclusions are usually classified as granules.

- *Granule inclusion.* Membrane-free and densely packed, this type of inclusion has many granules each containing specific substances. For example, *polyphosphate granules*, also known by the names *metachromatic granules* and *volutin*, have granules of polyphosphate that are used to synthesize ATP and are involved in other metabolic processes. A polyphosphate granule appears red under a microscope when stained with methylene blue.

- *Vesicle inclusion.* This is a protein membrane inclusion commonly found in aquatic photosynthetic bacteria and cyanobacteria such as *phytoplankton*, which suspends freely in water. These bacteria use vesicle inclusions to store gas that give the cell buoyancy to float at a depth where light, carbon dioxide (CO_2), and nutrients—all required for photosynthesis—are available.

Eukaryotic Cells

A *eukaryotic cell* (Fig. 4-6) is larger and more complex than a prokaryotic cell and found in animals, plants, algae, fungi, and protozoa. When you look at a eukaryotic cell with a microscope you'll notice a highly organized structure of organelles that are bound by a membrane. Each organelle performs a specialized function for the cell's metabolism. Eukaryotic cells also contain a membrane-bound nucleus where the cell's DNA is organized into chromosomes.

Depending on the organism, a eukaryotic cell may contain external projections called flagella and cilia. These projections are used for moving substances along the cell's surface or for moving the entire cell. *Flagella* move the cell in a wavelike motion within its environment. *Cilia* move substances along the cell's

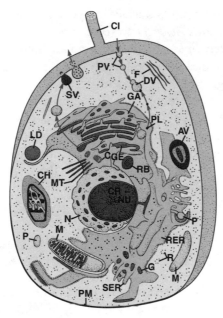

Fig. 4-6. A eukaryotic cell.

surface and also aid in movement of the cell. Flagella and cilia are comprised of axoneme microtubules. An *axoneme microtubule* is a long, hollow tube made of protein called a *tubulin*.

CELL WALL

Many eukaryotic cells have a cell wall. The composition of the cell wall differs with each organism. For example, the cell walls of many fungi are composed of chitin cellulose. *Chitin* is a polysaccharide, which is a polymer of N-acetylglucosamine (NAG) units. The cell wall of other fungi is made of *cellulose*, which is also a polysaccharide. Cellulose is also found in the cell wall of plants and many algae. Yeast has a cell wall composed of *glucan* and *mannan*, which are two polysaccharides.

In contrast, protozoa have no cell wall and instead have a pellicle. A *pellicle* is a flexible, proteinaceous covering. Eukaryotic cells of other organisms (such as animals) that lack a cell wall have an outer plasma membrane that serves as an outside cover for the cell. The outer plasma membrane has a sticky carbohydrate called *glycocalyx* on its surface. Glycocalyx is made up of covalently bonded lipids and proteins in order to form glycolipid and glycoprotein in the plasma membrane. *Glycolipid* and *glycoprotein* anchor the glycocalyx to the cell, giving the cell strength and helping the cell to adhere to other cells. Glycocalyx is also a molecular signature used to identify the cell to other cells. White blood cells use this to identify a foreign cell before destroying it.

A eukaryotic cell lacks peptidoglycan, which is critical in fighting bacteria with antibiotics. A bacterium is a prokaryotic cell. *Peptidoglycan* is the framework of a prokaryotic cell's cell wall. Antibiotics such as penicillin attack peptidoglycan resulting in the destruction of the cell wall of a bacterium. Eukaryotic cells invaded by the bacterium remain unaffected because eukaryotic cells lack peptidoglycan.

PLASMA MEMBRANE

The plasma membrane is a selectively permeable membrane enclosing the cytoplasm of a cell. This is the outer layer in animal cells. Other organisms have a cell wall as the outer layer and the plasma membrane is between the cell wall and the cell's cytoplasm. The cell wall is the outer covering of most bacteria, algae, fungi, and plant cells. In eubacterium, which is a prokaryotic microorganism, the cell wall contains peptidoglycan.

The plasma membrane surrounds a eukaryotic cell and serves as a barrier between the inner cell and its environment. The cytoplasmic membrane is composed of proteins and lipids. Carbohydrates are used to uniquely identify the cell to other cells. Lipids, known as sterols, help prevent the destruction of the cell when there is an increase in osmotic pressure and are mainly used for stability. *Lysis* is the destruction of a cell. Prokaryotic cytoplasm lacks certain features that are found in eukaryotic cytoplasm, such as a cytoskeleton. In a eukaryotic microorganism, the cytoskeleton provides support and shape for cells and helps transport substances through the cell.

The plasma membrane of a eukaryotic cell functions like the plasma membrane of a prokaryotic cell, which you learned about previously in this chapter. That is, substances enter and leave the cell through the cytoplasmic membrane by using simple diffusion, facilitated diffusion, osmosis, and active transport.

Eukaryotic cells extend parts or sections of plasma membrane. The extensions of the plasma membrane are called pseudopods. The word *pseudopod* means "false foot," and these "feet" enable the cell to have amoeboid motion. An *amoeboid motion* consists of muscle-like contractions that move the cell over a surface. Pseudopods are used to engulf substances and bring them into the cell, which is called endocytosis (a type of active transport). There are two types of endocytosis. These are phagocytosis (eat) and pinocytosis (drink). In *phago-cytosis,* solid particles are engulfed by the cell. An example is when a white blood cell engulfs and destroys a bacteria cell. In *pinocytosis,* liquid particles are brought into the cell. An example is when extracellular fluid containing a sub-stance is destroyed by the cell.

CYTOPLASM AND NUCLEUS

The cytoplasm of a eukaryotic cell contains cytosol, organelles, and inclusions, which is similar to the cytoplasm of the prokaryotic cell. Eukaryotic cytoplasm also contains a cytoskeleton that gives structure and shape to the cell and assists in transporting substances throughout the cell.

The nucleus of a eukarytoic cell contains DNA (hereditary information) and is contained within a nuclear envelope. DNA is also found in the mitochondria and chloroplasts. Depending on the organism, there can be one or more nucleoli within the nuclear envelope. A *nucleolus* (little nucleus) is the site of ribosomal RNA synthesis, which is necessary for ribosomes to function properly.

In the nucleus, the cell's DNA is combined to form several proteins called histones. The combination of about 165 pairs of DNA and nine molecular of his-tones make up the nucleosome. When a eukaryotic cell is not in the reproduc-tion phase, the DNA and its proteins look like a threaded mass called chromatin. When the cell goes through nuclear division, the strands of chromatin condense and coil together, producing rod-shaped bodies called chromosomes.

A eukaryotic cell uses a method of cell division during reproduction called mitosis. This is the formation of two daughter cells from a parent cell.

ENDOPLASMIC RETICULUM

The *endoplasmic reticulum* contributes to the mechanical support and distribu-tion of the cytoplasm and is the pathway for transporting lipids and proteins throughout the cell. The endoplasmic reticulum also provides the surface area for the chemical reaction that synthesizes lipids, it stores lipids and proteins until the cell needs them.

The endoplasmic reticulum consists of cisterns, which are a network of flat-tened membranous sacs. The end of these cisterns can be pinched off to become membrane-enclosed sacs called secretory vesicles. Vesicles transport synthe-sized material in the cell.

There are two kinds of endoplasmic reticula.

- *Rough endoplasmic reticulum.* Covered by ribosomes, which are the sites for synthesizing protein.
- *Smooth endoplasmic reticulum.* Not covered by ribosomes, this is the site for synthesizing lipids.

GOLGI COMPLEX

The Golgi complex is considered the "Fedex System" of the cell because it packages and delivers proteins, lipids, and enzymes throughout the cell and to the environment. The Golgi complex contains cisterns stacked on top of each other. A cistern is a sac or vessel and is filled with proteins or lipids (packaged), detached from the Golgi complex, and transported to another part of the cell.

LYSOSOME

A *lysosome* is a sphere in animal cells that is formed by, but is separate from, the Golgi complex, it contains enzymes used to digest molecules that have entered the cell. Think of lysosomes as the digestive system of the cell. For example, lysosomes in a white blood cell digest bacteria that is ingested by the cell during phagocytosis.

MITOCHONDRION

The *mitochondrion* is an organelle that is comprised of a series of folds called *cristae* that is responsible for the cell's energy production and cellular respiration. Chemical reactions occur within the center of the mitochondrion, called the *matrix*; it is filled with *semifluid* in which adenosine triphosphate (ATP) is produced. ATP is the energy molecule in the cell. The mitochondrion is the powerhouse of the cell.

CHLOROPLAST

Eukaryotic cells of green plants and algae contain *plastids,* one of which is chloroplast. *Chloroplasts* are organelles that contain pigments of chlorophyll and carotenoids used for gathering light and enzymes necessary for photosynthesis. *Photosynthesis* is the process that converts light energy into chemical energy. The pigment is stored in membranous sacs called *thylakoids* that are arranged in stacks called *grana*.

CENTRIOLE

A *centriole* is a pair of cylindrical structures near the nucleus that is comprised of microtubules and aids in the formation of flagella and cilia. The centriole also has a part in eukaryotic cell division.

Quiz

1. A gram-positive cell wall has
   (a) many layers of peptidoglygan, which repels the crystal of violet dye when the cell is stained
   (b) many layers of peptidoglygan, which retains the crystal of violet dye when the cell is stained
   (c) one layer of peptidoglygan, which retains the crystal of violet dye when the cell is stained
   (d) one layer of peptidoglygan, which repels the crystal of violet dye when the cell is stained

2. A cytoplasmic membrane is
   (a) a membrane that provides a selective barrier between the nucleus and the cell's internal structures
   (b) the cell wall
   (c) a membrane that provides a barrier between the cell's internal structures
   (d) a membrane that provides a selective barrier between the cell wall and the cell's internal structures

3. Amphipathic means
   (a) that the cytoplasmic membrane of a cell is bilayered and contains both polar and nonpolar parts
   (b) that the cytoplasmic membrane of a cell is unilayered and contains polar parts
   (c) that the cytoplasmic membrane of a cell is unilayered and contains nonpolar parts
   (d) that the cytoplasmic membrane of a cell is resistant to all substances residing outside the cell

4. What is the function of the nonpolar part of the cytoplasmic membrane?

hydrophobic

    (a) The cytoplasmic membrane does not have a nonpolar part.
    (b) The nonpolar part of the cytoplasmic membrane prevents extracellular fluid from leaving the cell and intracellular fluid from entering the cell.
    (c) The nonpolar part of the cytoplasmic membrane is hydrophobic and prevents extracellular fluid from entering the cell and intracellular fluid from exiting the cell.
    (d) The nonpolar part of the cytoplasmic membrane is to position organelles within the cell.

5. The function of the transport protein is
    (a) to regulate cell division
    (b) to regulate the positioning of organelles within the cell
    (c) to regulate movement of molecules through the cytoplasmic membrane
    (d) to give a cell its color

6. The function of the channel protein is
    (a) to direct movement of a cell through channels in its environment
    (b) to channel substances among organelles within the cell
    (c) to form pores (called channels) in the cytoplasmic membrane that permit the flow of molecules through the cytoplasmic membrane
    (d) to form pores (called channels) in the nucleus membrane that permit the flow of molecules through the nuclei of the cell

7. Passive transport is
    (a) the process of moving substances through the cytoplasmic membrane without expending energy by using a concentration gradient
    (b) the process of moving substances through the cytoplasmic membrane without expending energy by using a transport protein
    (c) the process of moving substances through the cytoplasmic membrane by expending energy by using a concentration gradient
    (d) the process of moving substances through the nucleus membrane without expending energy by using a concentration gradient

8. What is facilitated diffusion?
    (a) A passive transport process in which molecules or ions of a substance move from a region of lower concentration to a region of higher concentration without the assistance of an integral protein
    (b) A passive transport process in which molecules or ions of a substance move from a region of lower concentration to a region of higher concentration with the assistance of an integral protein

   (c) An active transport process in which molecules or ions of a substance move from a region of higher concentration to a region of lower concentration with the assistance of an integral protein

   (d) A passive transport process in which molecules or ions of a substance move from a region of higher concentration to a region of lower concentration with the assistance of an integral protein

9. Active transport is
   (a) the movement of a substance across the cytoplasmic membrane in the direction of the gradient by using energy provided by the cell
   (b) the formation of pores (called channels) in the cytoplasmic membrane that permit the flow of molecules through the cytoplasmic membrane
   (c) the movement of a substance across the cytoplasmic membrane against the gradient by using energy provided by the cell
   (d) is a process in which molecules or ions of a substance move from a region of higher concentration to a region of lower concentration with the assistance of an integral protein

10. What are the two types of endocytosis?
   (a) Facilitated diffusion and passive transport
   (b) Phagocytosis and pinocytosis
   (c) Hydrolysis and cytosol
   (d) Nucleoid and cytosol

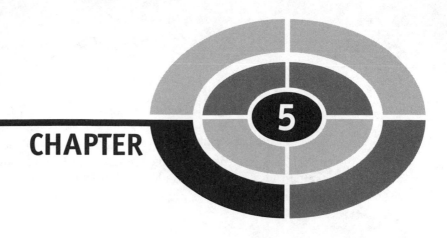

CHAPTER 5

The Chemical Metabolism

Those of us who need to shed a few pounds tend to blame our weight gain on our slow metabolism. *Metabolism* is the collection of biochemical reactions that happen in our bodies; an example would be the series of reactions that occur when our digestive system breaks down the food that we eat to energy. All living organisms have a metabolism, including microorganisms. In this chapter you will learn about the metabolism of the smallest living part of any organism—the cell.

Riding the Metabolism Cycle

The components of a cell, including its plasma membrane and cell wall (and organelles in eukaryotic organisms), are composed of macromolecules that are linked together to form these structures. The macromolecules are assembled from building blocks called *precursor metabolites*. Think of precursor metabolites as the bricks that are used to build a wall and the wall as the macromolecule. With the energy from ATP (adenosine triphosphate), these precursor metabolites are used to construct or build larger molecules. ATP is the short term energy storage molecule of the cell. Think of it as the battery pack of the cell. Cells use energy from ATP and enzymes to connect smaller molecules to form macromolecules. The cell grows as macromolecules are linked together and continue to grow into cellular structures such as organelles, plasma membranes, and cell walls.

Catabolic and Anabolic: The Only Reactions You Need

A biochemical reaction is called a metabolic reaction. Metabolic reactions fall into one of two classifications. These are *catabolic* reactions (*catabolism)* and *anabolic* reactions (*anabolism).*

A *catabolic reaction* is a metabolic reaction that releases energy as large molecules that are broken down (metabolized) into small molecules. An example is when triglycerides and diglycerides are metabolized into glycerol and fatty acids.

An *anabolic reaction* requires energy as small molecules are combined to form large molecules. This type of reaction is called *endergonic* because it uses free energy. For example, an anabolic reaction is the synthesis of phospholipids from glycerol and fatty acids in order to build the cell plasma membrane.

A Little Give and Take: Oxidation-Reduction

Metabolic reactions sometime involve the transfer of electrons from one molecule to another. One molecule *donates* an electron and another molecule *accepts* the electron. This transfer of electrons is called *oxidation-reduction* or *redox reaction*. A redox reaction is comprised of two events. The first event happens when a molecule donates an electron. This is called *oxidation*. The second event happens when another molecule accepts the donated electron. This is called *reduction*.

The cell uses *electron carrier molecules* to carry electrons between areas within the cell. Think of these carrier molecules as "shuttle buses." Carrier molecules are necessary because the cytoplasm of the cell does not contain free electrons.

Two important electron carrier molecules that are used in cell metabolism are

- nicotinamide adenine dinucleotide (NAD+)
- flavine adenine dinucleotide (FAD)
- nicotinamide adenine dinucleotide phosphate (NADP+)

For example, when synthesizing ATP, NAD+ carries electrons of a hydrogen (H) atom, making NADH. FAD carries two electrons of hydrogen making $FADH_2$. Very often electrons of hydrogen atoms are the electrons transported by the carrier molecule. NADP+ is used to reduce CO_2 to carbohydrates during the *Dark Phase* of photosynthesis.

Making Power: ATP Production

When enzymes break down nutrients (larger molecules) into smaller molecules, the energy that is released can be stored and used for future anabolic reactions. Here are the steps to form ATP:

1. **Substrate-level phosphorylation:** Phosphate is transferred from another phosphorylated organic compound to ADP to make ATP during an exergonic reaction.

2. **Oxidative phosphorylation:** Energy from redox reactions of biochemical respiration is used to attach an inorganic phosphate to ADP to make ATP.

3. **Photophosphorylation:** Energy from sunlight is used to phosphorylate ADP with inorganic phosphate.

What's Your Name: Naming and Classifying Enzymes

Enzymes are named according to the substrate that they act upon, and most end in the suffix "-ase." Enzymes are classified into six major groups based on their actions. These classifications are:

- **Hydrolases:** Enzymes in the *hydrolases* group increase a catabolic reaction by introducing water into the reaction. This reaction is called *hydrolysis*. For example, *lipase* (lipid + ase) is an enzyme that is used to break down lipid molecules.

- Isomerases: Enzymes in the *isomerase* group rearrange atoms within the substrate rather than add or subtract anything from the reaction. Phospho-glucoisomerase is an example of an isomerase because it converts glucose 6-phosphate into fructose 6-phosphate during the breakdown of glucose.

- Ligases: These are anabolic reactions. These enzymes join molecules together and use energy in the form of ATP. An example is DNA ligase to synthesis DNA.

- Lyases: Enzymes in the *lyases* group split molecules without using water in a catabolic reaction. For example, 1,6-biphosphate aldolase splits fructose 1,6 biphosphate into G-3P and DHAP during glycolysis. These are anabolic reactions.

- Oxidoreductases: Enzymes in the *oxidoreductases* group oxidize (remove) electrons or reduce (add) electrons to a substrate in both catabolic and anabolic reactions. An example is lactic acid dehydrogenase, which oxidizes pyruvate to form lactic acid during fermentation.

- Transferases: Enzymes in the *tranferases* group transfer functional groups from one molecule or another substrate in an anabolic reaction. A functional group could be amino acids, a phosphate group, or an acetyl group. For example, hexokinase transfers a phosphate group from ATP to glucose in the first step in the breakdown of glucose during the process of gycolysis.

Brewing Up Protein

Most enzymes are proteins that can be inactive or active. An *inactive enzyme* does not act as a catalyst to increase the speed of a metabolic reaction. An *active enzyme* is a catalyst. An inactive enzyme is composed of *apoenzyme*; when an apoenzyme binds to its cofactor the enzyme becomes active and is called a *holoenzyme*.

A *cofactor* is a substance that is either an inorganic ion, such as iron, magnesium, or zinc, or an organic molecule. Organic cofactors are called coenzymes. A coenzyme is a molecule that is required for metabolism. NAD, NADP, and FAD are examples of coenzymes. Some vitamins are coenzyme precursors.

The Magic of Enzymes: Enzyme Activities

All chemical reactions including those that occur in metabolism, need a boost of energy to get started. The energy needed to begin a chemical reaction is called *activation energy*. An enzyme catalyzes a reaction by lowering the activation energy. Heat can lower the activation energy and set off a reaction. However, the temperature would be so high that the cell would die before the activation energy threshold could be reached.

Enzymes are needed for metabolism to occur in a timely fashion. The activity of enzymes depends upon how closely their functional sites fit with their substrates. The shape of the enzyme's functional site is called its *active site*. This site fits in regard to the shape of the substrate. The active site of the enzyme compliments the shape of its substrate. A perfect fit *does not* occur until the substrate and enzyme bind together to form an enzyme-substrate complex.

PH

The chemical denaturing of enzymes is caused by very high or very low pH. H^+ ions that are released from acids and accepted by bases interfere with hydrogen bonding. If we change the pH of the environment of unwanted microorganisms, we can control their growth by denaturing their *proteins*. An example is vinegar, which is acetic acid; it has a pH of 3.0. Vinegar acts as a preservative in " pickling" vegetables. Ammonia has a pH of 11. Ammonia is a base, and for this reason we use ammonia as a cleaner and disinfectant.

ENZYME SUBSTRATE CONCENTRATION

As substrate concentrations increase, enzyme activity also increases. When all enzyme binding sites have bound to a substrate, the enzymes have reached their *saturation point*. If more substrate is added, the rate of enzyme activity *will not* increase. One way organisms regulate their metabolism is by controlling the quantity and timing of enzyme synthesis.

The Right Influences: Factors Affecting Enzymes

The ability of an enzyme to lower the activation required for metabolism is influenced by three factors. These are pH, temperature, and the concentration of enzyme, substrate, and product.

Temperature

Changes in temperature change the shape of the active site, and therefore, influence the fit between the active site and the substrate. Enzymes in humans work best at about 37 degrees Celsius. This is the same temperature at which enzymes work best for some pathogenic microorganisms, too. Once the temperature reaches the point that radically changes the shape of the active site, the bond between the active site and the substrate is broken and makes the enzyme ineffective. This is called *thermal denaturation*. Denatured enzymes lose their specific three-dimensional shape, making them nonfunctional. For example, the clear liquid portion of an egg turns to a white solid when the egg is heated. The clear liquid is made up of proteins. Heating these proteins denatures them.

Inhibitors

There are substrates that block active sites from bonding to a substrate. These substances are called *inhibitors*. There are two kinds of inhibitors: competitive and noncompetitive. A *competitive inhibitor* is a substance that binds to the active site of an enzyme, thus preventing the active site from binding with the substrate. For example, sulfa drugs contain the chemical sulfanilamid. Sulfa drugs inhibit microbial growth by fitting into the active site of an enzyme required in the conversion of paraaminobenzoic acid (PABA) into the B vitamin folic acid. Folic acid is needed for DNA synthesis in bacteria and thus prevents bacteria from growing. A noncompetitive inhibitor binds to another site on the enzyme called the allosteric site and in doing so alters the shape of the active site of the enzyme. The shape of the active site no longer complements the corresponding site on the substrate and therefore no binding occurs. Noncompetitive inhibitors do not bind to active sites.

CARBOHYDRATE METABOLISM

Carbohydrates are the main energy source for metabolic reactions and glucose is the most used carbohydrate in metabolism. Energy is produced by breaking down (catabolized) glucose in a process called *glycolysis*, which takes place in the cytoplasm of most cells. Glycolysis, also known as the Embden-Meyerhof pathway, is the oxidation of glucose to pyruvic acid. In glycolysis, which originated from the Greek word *glykys* meaning "sweet" and *lysein* meaning "loosen," enzymes split a six-carbon sugar into two three-carbon sugars, which are then oxidized. Oxidation releases energy and rearranges atoms to form two molecules of pyruvic acid. It is during this process that NAD^+ is reduced to NADH with a net production of two ATP molecules.

In the presence of O_2 (aerobic environment), pyruvic acid enters the bridging pathway and becomes connected to acetyl CoA. It then enters the Krebs cycle, which will result in the production of three NADH and, one $FADH_2$ molecules. In the absence of O_2 (anaerobic environment), the NADH produced during glycolysis is oxidized and an organic compound accepts the electrons. This process is called *fermentation*. This pathway of fermentation results in fewer ATP molecules.

Embden-Meyerhof Pathway

The Embden-Meyerhof pathway (Fig. 5-1) is used by some bacteria to catabolize glucose to pyruvic acid. For example, *Pseudomonas aeruginosa*, which is the bacteria that infects burn victims, uses the Embden-Meyerhof pathway. The Embden-Meyerhof pathway yields one molecule of ATP.

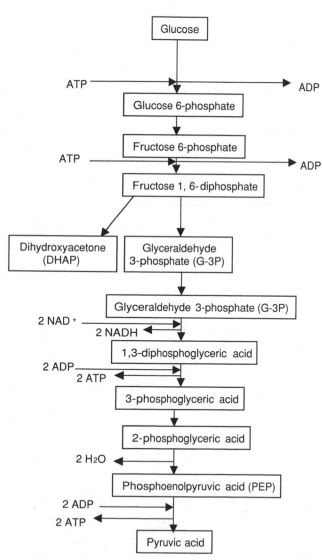

Fig. 5-1.   Diagram of glycolysis: Embden-Meyerhoff pathway.

Pyruvic Acid

Before entering the Krebs cycle, the pyruvic acid produced from the breakdown of glucose must be further processed by converting it to acetyl-coenzyme A. This is accomplished by the enzyme complex *pyruvate dehydrogenase*. CO_2 is removed from pyruvic acid and the product is an acetyl group (a two carbon group). The acetyl group is attached to coenxzyme A and the product is called acetyl-CoA. The removal of CO_2 is called decarboxylation.

The Krebs Cycle

The *Krebs cycle* (Fig, 5-2) is a series of biochemical reactions that occur in the mitochondria of eukaryotic cells and in the cytoplasm of prokaryotes. The Krebs

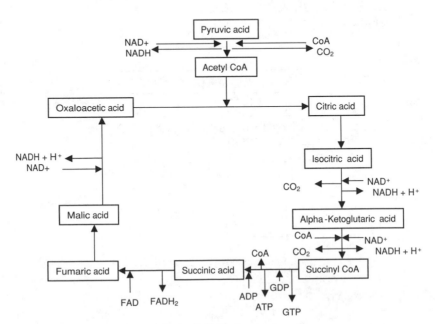

Fig. 5-2.   The Krebs Cycle: Citric Acid Cycle

cycle, also known as the citric acid cycle and the tricarboxylic acid (TCA) cycle, is named for Sir Hans Krebs, a biochemist, who in the 1940s explained how these reactions work. In the Krebs cycle, acetyl-CoA is split into carbon dioxide and hydrogen atoms.

The carbon dioxide diffuses out of the mitochondria in eukaryotic cells and eventually out of the cell itself. This series of reactions is called a cycle because as one acetyl group enters the Krebs cycle and is metabolized, oxaloacetate combines with another acetyl group to form citric acid and coenzyme A, which go through the cycle again. As each acetyl group goes through the cycle, two molecules of CO_2 are formed from the oxidation of its two carbon atoms.

Three pairs of electrons are transferred to NAD and one pair to FAD. These coenzymes are important because they carry large amounts of energy. For every molecule of acetyl-CoA that enters the Krebs Cycle, a molecule of ATP is produced. The Krebs cycle also provides substances for bacteria and other prokaryotic cellular activities.

THE NEW CHAIN GANG:
THE ELECTRON TRANSPORT CHAIN

Glycolysis, the bridging reaction, and the Krebs cycle result in the synthesis of only four ATP molecules when one glucose is oxidized to six CO_2 molecules. Most of the ATP that is generated comes from the oxidation of NADH and $FADH_2$ in the electron transport chain.

The electron transport chain, which occurs in the mitochondria in eukaryotic cells and in the cytoplasm of prokaryotic cells, is composed of a series of electron carriers that transfer electrons from donor molecules, such as NADH and $FADH_2$ to an acceptor atom like O_2. The electrons move down an energy gradient, like water flowing down a series of waterfalls in rapids.

The difference in free energy that occurs between O_2 and NADH releases large amounts of energy. The energy changes that occur at several points in the chain are very large and can provide the eventual production of large amounts of ATP. The free energy that electrons have entering the electron transport chain is greater in the beginning than at the end. It is this energy that enables the protons (H^+) to be pumped out of the mitochondrial matrix.

When the electrons move through the chain they transfer this energy to the pumps within the plasma membrane. The electron transport chain will separate the energy that is released into smaller sections, or steps. The reactions of the electron chain take place in the inner membrane of the mitochondria in eukaryotic cells or in the plasma membrane in prokaryotic cells. In the mitochondria this system is set up into four complexes of carriers.

Each of these carriers transports electrons part of the way to O_2 (which is the final electron acceptor). The carriers, coenzyme Q and cytochrome C, connect these complexes. This process by which energy comes from the electron transport chain is provided by protons (H^+) and are used to make ATP.

Three ATP molecules can be synthesized from ADP and Pi when two electrons pass from NADH to an atom of O_2.

The electron transport chain used by bacteria and other prokaryotes can differ from the mitochondrial chain used in eukaryotic organisms. Bacteria, for example, vary in their electron carriers. Bacteria use *cytochromes*, heme proteins that carry electrons through the electron transport chain. (A *heme* is an organic compound, the center of which contains an iron atom surrounded by four nitrogen atoms.) Electrons can enter at several points and leave through terminal oxidases. Prokaryotic and eukaryotic electrons work using the same fundamental principles, although they differ in construction.

The electron transport chain in *E. coli* bacteria, for example, transports electrons from NADH to acceptors and moves protons across the plasma membrane. The *E. coli* electron transport chain is branched and contains different cytochromes. The two branches are cytochrome d and cytochrome o. Coenzyme Q donates electrons to both branches. These chains operate in different conditions. For example, the cytochrome d branch will function when O_2 levels are low and does not actively pump protons, whereas the cytochrome o branch operates in higher O_2 concentrations and is a proton pump.

During the aerobic metabolism of a single glucose molecule, ten pairs of electrons from NAD produce thirty ATP molecules, and two pairs of FAD produce four ATP molecules, making a total of 34 ATP molecules. Four substrate-level ATPs make a total of 38 molecules of ATP from one molecule of glucose.

The energy captured occurs through a process called *chemiosmosis,* formulated by British biochemist Peter Mitchell, who won the Nobel Prize in 1978. In chemiosmosis electrons flow down their electrochemical gradient across the inner mitochondrial membrane in eukaryotes and the cell membrane in prokaryotes through ATP syntase.

If the organism is in an aerobic environment, there are enzymes that can break down harmful chemicals. An example of such a chemical is hydrogen peroxide (H_2O_2). If the organism is in an anaerobic environment, they do not possess or cannot produce these aerobic enzymes and are susceptible to damage by O_2. An example is the free radical superoxide. Organisms that follow this pathway produce less ATP. An example of these types of organisms is *lactobacillus*.

Fermentation

Fermentation is the partial oxidation of glucose or another organic compound to release energy; it uses an organic molecule as an electron acceptor rather than an electron transport chain. In a simple fermentation reaction, NADH reduces pyruvic acid from glycolysis to form lactic acid. Another example involves two reactions. The first is a decarboxylation reaction where CO_2 is given off (this is the CO_2 that causes bakery items to rise), followed by a subsequent reduction reaction that produces ethanol.

The essential function of fermentation is the regeneration of NAD^+ for glycolysis so ADP molecules can be phosphorylated to ATP. The benefit of fermentation is that it allows ATP production to continue in the absence of O_2. Microorganisms that ferment can grow and colonize in an anaerobic environment.

Microorganisms produce a variety of fermentation products. The products of fermentation of cells are waste products of the cells, but many are useful to humans. These include *ethanol* (the alcohol that humans can drink, like in beer, wine, and liquor), *acetic acid* (vinegar), and *lactic acid* (found in cheese, sauerkraut, and pickles). Other fermentation products are very harmful to humans. An example is the bacterium *Clostridium perfringens,* which ferments hydrogen and is associated with gas gangrene. Gangrene is involved in necrosis or the "death" of muscle tissue.

Other Catabolic Pathways

There are two other important catabolic pathways. These are lipid catabolism and protein catabolism. Both of these convert substances (lipids and proteins) into ATP, providing the microorganism with energy.

LIPID CATABOLISM

Lipids, including fats, which consist of glycerol and fatty acids, can be involved in ATP production. Enzymes called *lipases* hydrolyze the bonds attaching the glycerol to the fatty acid chains. Glycerol is converted to dihydroxyacetone phosphate (DHAP), which is oxidized to pyruvic acid in glycolysis. The fatty acids are broken down in catabolic reactions called *beta-oxidation*. In beta-oxidation,

enzymes split off pairs of hydrogenated carbon atoms that make up the fatty acids. The enzymes then join these pairs to coenzyme A to form acetyl-CoA. This happens until the entire fatty acid has been converted to molecules of acetyl-CoA. Acetyl-CoA is utilized in the cycle to generate ATP.

PROTEIN CATABOLISM

Most organisms break down proteins only when glucose and fats are unavailable. Some bacteria that spoil food, pathogenic bacteria, and fungi normally catabolize proteins as energy sources.

Proteins are too large to cross the cell membrane. Microorganisms secrete an enzyme called *protease*, which splits the protein into amino acids outside the cell. Amino acids are then transported into the cell, where specialized enzymes split off amino groups in reactions called *deamination*. These molecules then enter the Krebs cycle.

These are examples of how the catabolism or breakdown of proteins, carbohydrates, and lipids can be sources of electrons and proteins during cellular respiration. The pathways of glycolysis and the Krebs cycle are catabolic roadways or tunnels where high-energy electrons from these organic molecules can flow through on their energy-releasing journey.

Photosynthesis

Some organisms use anabolic pathways to synthesize organic molecules from inorganic carbon dioxide. Most of these phototrophic organisms are autotrophic and are capable of surviving and growing on carbon dioxide as their only carbon source. The energy from sunlight is used to reduce CO_2 to carbohydrates. This process is called photosynthesis. The ability of an organism to "photosynthesize" depends on the presence of light-sensitive pigments, called chlorophyll, or related compounds. These pigments are found in plants, algae, and certain bacteria.

The growth of these photosynthetic organisms can be explained by two separate types of reactions. In *light reactions*, light energy is converted into chemical energy and *dark reactions* in which the chemical energy from the light reactions is used to reduce carbon dioxide (CO_2) to carbohydrates. For the growth and survival of autotrophic organisms, energy is supplied in the form of an ATP molecule. Electrons for the reduction of CO_2 come from NADPH.

NADPH is produced from the reduction of $NADP^+$ by electrons from electron donors such as water.

Purple and green bacteria use light most of the time to form ATP. They produce NADPH from reducing materials present in their environment, such as reduced sulfur compounds (H_2S), organic compounds, as photosynthetic electron donors for CO_2 fixation. Green plants, algae, and cyanobacteria do not use reduced sulfur compounds or organic compounds to obtain reducing power. Instead, they obtain electrons for $NADP^+$ reduction by splitting water molecules. By splitting water, oxygen (O_2) is produced as a byproduct. The reduction of $NADP^+$ to NADPH by these organisms is dependent on light and therefore a light-mediated event. Due to the production of molecular oxygen (O_2) the process of photosynthesis in these organisms is called *oxygenic photosynthesis*. In contrast, the purple and green bacteria do not produce oxygen. This process is called *anoxygenic photosynthesis*.

These photosynthetic organisms capture the light with pigment molecules. An important pigment molecule is chlorophyll. There are different structures of chlorophyll, the most common of which are *chlorophyll a* and *chlorophyll b*. Chlorophyll a is the principal chlorophyll of higher plants, most algae, and cyanobacteria. Purple and green bacteria have chlorophylls of a different structure, called *bacteriochlorophyll*.

Accessory pigments, such as *carotenoids* and *phycobilins,* are also involved in capturing light energy. Carotenoids play a photoprotective role, preventing photooxidative damage to the phototrophic cell. Phycobilins serve as light-harvesting pigments.

Cells arrange numerous molecules of chlorophyll and accessory pigments within membrane systems called photosynthetic membranes. The location of these membranes differs between eukaryotic and prokaryotic microorganisms. In eukaryotic organisms, photosynthesis occurs in specialized organelles called *chloroplasts*. The chlorophyll pigments are attached to "sheet-like" membrane structures of the chloroplasts. These photosynthetic membrane structures are called *thylakoids*. These thylakoids are arranged in stacks called *grana*. Thylakoids resemble stacks of pennies. Each stack is called a *granum*.

In prokaryotic organisms, there are no chloroplasts. The photosynthetic pigments are integrated in a membrane system that arises from the cytoplasmic membranes.

PHOTOSYSTEMS I AND II

Electron flow in oxygenic phototrophs involves two sets of photochemical reactions. Oxygenic phototrophs use light energy to generate ATP and

NADPH. The electrons from NADPH come from the splitting of H_2O to get oxygen (O_2) and electrons (H^+). The two systems of light reactions are called *photosystem I* and *photosystem II*. Photosystems I and II function together in the oxygenic process. Under certain conditions many algae and some cyanobacteria can carry out cyclic photophosphorylation using only photosystem I. These organisms can obtain reducing power from sources other than water. This requires the presence of anaerobic conditions and a reducing substance, such as H_2 or H_2S. Photosystem II is responsible for splitting water to yield $H_2 + O$.

Dark Phase Reactions

Many photosynthetic bacteria contain carboxysomes in their cytoplasm. These carbosomes contain many copies of the complex enzyme *Rubisco* (the most abundant and probably the most important enzyme on the planet) to start the Calvin Cycle.

The fixation of CO_2 by most photosynthetic and autotrophic organisms involves the biochemical pathway called the *Calvin Cycle*. The Calvin Cycle is a reductive, energy-demanding process in which reducing equivalents from NADPH and energy from ATP are used to reduce CO_2 to small carbohydrates that are metabolized to glucose and ultimately to more complex carbohydrates such as starch, sucrose and glycogen. Other substances with carbon can be used but CO_2 is preferred.

Quiz

1. What is the energy storage molecule?
   (a) ATP
   (b) Adenosine oxidate
   (c) Phospholipids
   (d) NAAD

2. What reaction releases energy as large molecules are broken down into small molecules?
   (a) Anabolic reaction
   (b) Catabolic reaction

    (c) Anabolism
    (d) Hydrolase reaction

3. What reaction combines small molecules to form large molecules?
    (a) Anabolic reaction
    (b) Catabolic reaction
    (c) Anabolism
    (d) Hydrolase reaction

4. The energy needed to begin a chemical reaction is called?
    (a) Metabolism
    (b) Activation energy
    (c) Saturation point
    (d) Inhibitors

5. What process is used to break down glucose?
    (a) Anabolic reaction
    (b) Catabolic reaction
    (c) Anabolism
    (d) Hydrolase reaction

6. What pathway is used by some bacteria to catabolize glucose to pyruvic acid?
    (a) Pyruvic acid pathway
    (b) Enter-Doudoroff pathway
    (c) *Pseudomonas* pathway
    (d) *Aeruginosa* pathway

7. What is acetyl-CoA split into in the Krebs cycle?
    (a) Hydrogen and oxygen
    (b) Oxygen and carbon
    (c) Carbon dioxide and hydrogen
    (d) Carbon and hydrogen

8. What process partially oxidizes sugar to release energy using an organic molecule as an electron acceptor?
    (a) Fermentation
    (b) Oxidation
    (c) Oxidation reduction
    (d) Lactation

9. What is caused when there is a chemical denaturing of enzymes?
   (a) Protein creation
   (b) Lipid creation
   (c) Very high or low pH
   (d) Saturation point is reached

10. What blocks active sites from bonding to a substrate?
   (a) Carbohydrates
   (b) Temperature
   (c) Inhibitors
   (d) Pyruvic acid

CHAPTER

Microbial Growth and Controlling Microbial Growth

Microorganisms use chemicals called nutrients for growth and development. They need these nutrients to build molecules and cellular structures. The most important nutrients are carbon, hydrogen, nitrogen, and oxygen. Microorganisms get their nutrients from sources in their environment. When these microorganisms obtain their nutrients by living on or in other organisms, they can cause disease in that organism by interfering with their host's nutrition, metabolism, and, thus disrupting their host's homeostasis, the steady state of an organism.

Organisms can be classified in two groups depending on how they feed themselves. Organisms that use carbon dioxide (CO_2) as their source of carbon are called *autotrophs*. These organisms "feed themselves," *auto-* meaning "self" and *-troph* meaning "nutrition." Autotrophs make organic compounds from CO_2 and do not feed on organic compounds from other organisms. Organisms that obtain carbon from organic nutrients like proteins, carbohydrate, amino acids, and fatty acids are called *heterotrophs*. Heterotrophic organisms acquire or feed on organic compounds from other organisms.

Organisms can also be categorized according to whether they use chemicals or light as a source of energy. Organisms that acquire energy from redox reactions involving inorganic and organic chemicals are called *chemotrophs*. Organisms that use light as their energy source are called *phototrophs*.

Chemical Requirements for Microbial Growth

A microorganism requires two chemical elements in order to grow. These elements are carbon and oxygen.

CARBON

Carbon is one of the most important requirements for microbial growth. Carbon is the backbone of living matter. Some organisms, such as photoautotrophs, get carbon from carbon dioxide (CO_2).

OXYGEN

Microorganisms that use oxygen produce more energy from nutrients than microorganisms that do not use oxygen. These organisms that require oxygen are called *obligate aerobes*. Oxygen is essential for obligate aerobes because it serves as a final electron acceptor in the electron transport chain, which produces most of the ATP in these organisms. An example of an obligate aerobe is *micrococcus*. Some organisms can use oxygen when it is present, but can continue to grow by using fermentation or anaerobic respiration when oxygen is not available. These organisms are called *facultative anaerobes*. An example of a facultative anaerobe is *E. coli* bacteria, which is found in the large intestine of vertebrates, such as humans.

Some bacteria cannot use molecular oxygen and can even be harmed by it. Examples include *Clostridium botulinum*, the bacterium that causes botulism, and *Clostridium tetani*, the bacterium that causes tetanus. These organisms are called *obligate anaerobes*. Molecular oxygen (O_2) is a poisonous gas to obligate anaerobes. Toxic forms of oxygen are:

- *Singlet oxygen*, which is in an extremely active, high-energy state. This is present in cells that use phagocytosis to ingest foreign bacteria cells.

- *Superoxide free radicals,* which is formed during normal respiration of organisms that use oxygen as a final electron acceptor. In obligate anaerobes, they form some superoxide free radicals in the presence of oxygen. These superoxide free radicals are incredibly toxic to cellular components. In order for organisms to grow in atmospheric oxygen they must produce the enzyme *superoxide dimutase* or (SOD). Superoxide dimutase neutralizes superoxide free radicals. Aerobic bacteria, facultative anaerobes growing aerobically, and aerotolerant anaerobes produce SOD, which converts the superoxide free radical into molecular oxygen (O_2) and hydrogen peroxide (H_2O_2). This can be seen when you place hydrogen peroxide on a wound infected with bacteria. When hydrogen peroxide is placed on the colony of bacterial cells that are producing catalase, oxygen bubbles are released. This is the "foaming" you see when you place hydrogen peroxide on a cut. Human cells also produce catalase, which converts hydrogen peroxide to water and oxygen.

- *Hydroxyl radical* (OH^-); this is a hydroxide ion. Most aerobic respiration produces some hydroxyl radicals. Aerotolerant anaerobes can tolerate oxygen, but cannot grow in an oxygen-rich environment. Aerotolerant bacteria ferment carbohydrates to lactic acid. Lactic acid inhibits the growth of aerobic competitors, establishing a good opportunity for the growth of these organisms. An example of a lactic acid–producing aerotolerant bacterium is *lactobacillus*, used in fermented food such as pickles and yogurt.

Culture Media

A *culture medium* is nutrient material prepared in the laboratory for the growth of microorganisms. Microorganisms that grow in size and number on a culture medium are referred to as a culture.

In order to use a culture medium must be *sterile*, meaning that it contains no living organisms. This is important because we only want microorganisms that we add to grow and reproduce, not others. We must have the proper nutrients, pH, moisture, and oxygen levels (or no oxygen) for a specific microorganism to grow.

Many culture media are available for microbial growth. Media are constantly being developed for the use of identification and isolation of bacteria in the research of food, water, and microbiology studies.

The most popular and widely used medium used in microbiology laboratories is the solidifying agent *agar*. Agar is a complex polysaccharide derived from red algae. Very few microorganisms can degrade agar, so it usually remains in a solid form. Agar media are usually contained in test tubes or Petri dishes. The test tubes are held at a slant and are allowed to solidify on an angle, called a *slant*. A slant increases the surface area for organism growth. When they solidify in a vertical tube it is called a deep. The shallow dishes with lids to prevent contamination are called Petri dishes. Petri dishes are named after their inventor, Julius Petri, who in 1887 first poured agar into glass dishes.

CHEMICALLY DEFINED MEDIA

For a medium to support microbial growth, it must provide an energy source, as well as carbon, nitrogen, sulfur, phosphorous, and any other organic growth factors that the organism cannot make itself, source for the microorganisms to utilize.

A *chemically defined* medium is one whose exact chemical composition is known. Chemically defined media *must* contain growth factors that serve as a source of energy and carbon. Chemically defined media are used for the growth of autotrophic bacteria. Heterotrophic bacteria and fungi are normally grown on *complex media*, which are made up of nutrients, such as yeasts, meat, plants, or proteins (the exact composition is not quite known and can vary with each mixture). In complex media, the energy, carbon, nitrogen, and sulfur needed for microbial growth are provided by protein. Proteins are large molecules that some microorganisms can use directly. Partial digestion by acids and enzymes break down proteins into smaller amino acids called *peptones*. Peptones are soluble products of protein hydrolysis. These small peptones can be digested by bacteria.

Different vitamins and organic growth factors can be provided by meat and yeast extracts. If a complex medium is in a liquid form it is called a *nutrient broth*. If agar is added, it is called a *nutrient agar*. Agar is not a nutrient; it is a solidifying agent.

ANAEROBIC GROWTH

Because anaerobic organisms can be killed when exposed to oxygen, they must be placed in a special medium called a *reducing medium*. Reducing media contain ingredients like sodium thioglycolate that attaches to dissolved oxygen and depletes the oxygen in the culture medium.

SELECTIVE AND DIFFERENTIAL MEDIA

In health clinics and hospitals, it is necessary to detect microorganisms that are associated with disease. Selective and differential media are therefore used. *Selective media* are made to encourage the growth of some bacteria while inhibiting others. An example of this is bismuth sulfite agar. Bismuth sulfite agar is used to isolate *Salmonella typhi* from fecal matter. *Salmonella typhi* is a gram-negative bacterium that causes salmonella. Differential media make it easy to distinguish colonies of desired organisms from nondesirable colonies growing on the same plate. Pure cultures of microorganisms have identifiable reactions with different media. An example is blood agar. Blood agar is a dark red/brown medium that contains red blood cells used to identify bacterial species that *destroy* red blood cells. An example of this type of bacterium is *streptococcus pyogenes*, the agent that causes strep throat.

MacConkey agar is both selective and differential. MacConkey agar contains bile salts and crystal violet, which inhibit the growth of gram-positive bacteria, and lactose, in which gram-negative bacteria can grow.

Enrichment cultures are usually liquids and provide nutrients and environmental conditions that provide for the growth of certain microorganisms, but not others.

PURE CULTURES

Infectious material or materials that contain pathogenic microorganisms can be located in pus, sputum, urine, feces, soil, water, and food. These infectious materials can contain several kinds if bacteria. If these materials are placed on a solid medium, colonies will form that are the exact copies of that same microorganism. A *colony* arises from a single spore, vegetative cell, or a group of the same organism that attaches to others like it into clumps or chains. *Microbial colonies* have distinct appearances that distinguish one microorganism from another.

The *streak plate method* is the most common way to get pure cultures of bacteria. A device called an *inoculating loop* is sterilized and dipped into a culture of a microorganism or microorganisms and then is "streaked" in a pattern over a nutrient medium. As the pattern is made, bacteria are rubbed off from the loop onto the nutrient medium. The last cells that are rubbed off the loop onto the medium are far enough apart to allow isolation of separate colonies of the original culture. (See Fig. 6-1.)

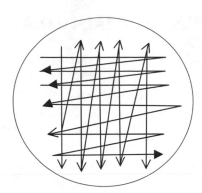

Fig. 6-1.   Streak plate method used to isolate bacteria.

PRESERVING BACTERIAL CULTURES

Two common methods of preserving microbial cultures are *deep freezing* and *lyophilization* or (*freeze drying*). Deep freezing is a process in which a pure culture of microorganisms is placed in a suspending liquid and frozen quickly at temperatures ranging from −50 to −95 degrees Celsius. With this type of freezing method, cultures can usually be thawed and used after several years. Lyophilization, or freeze drying, quick freezes suspended microorganisms at temperatures from −54 to −72 degrees Celsius while water is removed by using a high-pressure vacuum. While under the vacuum the container is sealed with a torch. The surrounding microbes in the sealed container can last for years. The organisms can be retained and revived by hydrating them and placing them into a liquid nutrient medium.

GROWING BACTERIAL CULTURES

Bacteria normally reproduce by a process called *binary fission*:

1. The cell elongates and chromosomal DNA is replicated.
2. The cell wall and cell membrane pinch inward and begin to divide.
3. The pinched parts of the cell wall meet, forming a cross wall completely around the divided DNA.
4. The cells separate into two individual cells.

Some bacteria reproduce by *budding*. A small outgrowth or bud emerges from the bacterium and enlarges until it reaches the size of the daughter cell. It then separates, forming two identical cells. Some bacteria, called filamentous bacteria (or actinomycetes), reproduce by producing chains of spores located at the tips of the filaments. The filaments fragment and these fragments initiate the growth of new cells.

GENERATION TIME

The generation time is the amount of time needed for a cell to divide. This varies among organisms and depends upon the environment they are in and the temperature of their environment. Some bacteria have a generation time of 24 hours, although the generation time of most bacteria is between 1 to 3 hours. Bacterial cells grow at an enormous rate. For example, with binary fission, bacteria can double every 20 minutes. In 30 generations of bacteria (10 hours), the number could reach one billion. It is difficult to graph population changes of this magnitude using arithmetic numbers, so logarithmic scales are used to graph bacterial growth.

Phases of Growth

There are four basic phases of growth: the lag phase, the log phase, the stationary phase, and the death phase.

THE LAG PHASE

In the lag phase there is little or no cell division. This phase can last from one hour to several days. Here the microbial population is involved in intense metabolic activity involving DNA and enzyme synthesis. This is like a factory "shutting down" for two weeks in the summer for renovations. New equipment is replacing old and employees are working, but no product is being turned out.

THE LOG PHASE

In the log phase, cells begin to divide and enter a period of growth or logarithmic increase. This is the time when cells are the most active metabolically. This is the time when the product of the factory must be produced in an efficient matter. In this phase, however, microorganisms are very sensitive to adverse conditions of their environment.

THE STATIONARY PHASE

This phase is one of equilibrium. The growth rate slows, the number of dead microorganisms equals the number of new microorganisms, and the population stabilizes. The metabolic activities of individual cells that survive will slow down. The reasons why the growth of the organisms stops is possibly that the nutrients have been used up, waste products have accumulated, and drastic harmful changes in the pH of the organisms environment have occurred.

There is a device called a *chemostat* that drains off old, used medium and adds fresh medium. In this way a population can be kept in the growth phase indefinitely.

THE DEATH PHASE

Here the number of dead cells exceeds the number of new cells. This phase continues until the population is diminished or dies out entirely.

Measurements of Microbial Growth

A *plate count* is the most common method of measuring bacterial growth. This method measures the number of viable cells. Plate counts may take 24 hours or more for visible colonies to form.

A plate count is performed by either a pour plate method or a special plate method. With the pour plate method either 1.0 milliliter or 0.1 milliliters of a bacterial solution is placed into a Petri dish. Melted nutrient agar is added, which is then gently agitated, or mixed. When the agar has solidified, the plate is then placed under incubation. With this technique, heat-sensitive microorganisms can be damaged by the melted agar and be unable to form colonies. To avoid death of cells due to heat, the spread plate method is mostly used. Here a 0.1-milliliter bacterial solution is added to the surface of a prepored, solid nutrient agar. The bacterial solution is then spread evenly over the medium.

When the amount of bacteria is very small, as in lakes and pure streams, bacteria can be counted by *filtration methods*. Here 100 milliliters of water are passed through a thin membrane, whose pores are too small for the bacteria to pass through. The bacteria that are retained from the filter are placed on a Petri dish containing a pad soaked with liquid nutrient. An example of bacteria that are grown using this method is coliform bacteria, which are indicators of fecal pollution of food or water.

When using the *direct microscopic count method*, a measured volume of bacteria is suspended in a liquid placed inside a designated area of a microscopic slide. For example, a 0.01-milliliter sample is spread over a marked square centimeter of a slide, stained, and viewed under the 0.1 immersion objective lens. The area for the viewing field is obtained. Once the number of bacteria is obtained or determined in several different fields, an average can be taken of the number of bacteria per viewing field. From this data, the number of bacteria in the square centimeter over which the sample has been spread can be calculated. Because the area on the slide contained 0.01 milliliters of sample, the number of bacteria in each milliliter of the suspension is the number of bacteria in the sample times 100.

Establishing Bacterial Numbers by Indirect Methods

Not all microbial cells must be counted to establish their number. With some types of work, estimating the turbidity is a practical way of monitoring microbial growth. Turbidity is the cloudiness of a liquid or the loss of transparency because of insoluble matter.

The instrument used to measure turbidity is a *spectrophotometer* or colorimeter. In the spectrophotometer, a beam of light is transmitted through the bacteria that are suspended in the liquid medium to a photoelectric cell. As bacteria growth increases, less light will reach the photoelectric cell. The change of light will register on the instrument's scale as the percentage of transmission. The amount of light striking the light-sensitive detector on the spectrophotometer is inversely proportional to the number of bacteria under normal cultures: the less light, the more bacteria.

Another indirect way of measuring bacterial growth is to measure the *metabolic activity* of the colony. In this method it is assumed that metabolic waste products, CO_2 (carbon dioxide) and acid, are in direct proportion to the number of bacteria present. The more bacteria we have, the more waste products we will also have.

For filamentous organisms, such as molds, a way to measure growth is by *dry weight*. The fungus is removed from its growth medium, filtered, and placed in a weighing bottle dried in a dessicator (a dessicator is a device that removes water). In bacteria the culture is removed from the medium by centrifugation.

Controlling Microbial Growth

It is very important to control microbial growth in surgical and hospital settings, as well as in industrial and food preparation facilities. There are many terms used to describe the fight to control microorganisms.

Sterilization is the destruction of all microorganisms and viruses, as well as endospores. Sterilization is used in preparing cultured media and canned foods. It is usually performed by steam under pressure, incineration, or a sterilizing gas such as ethylene oxide.

Antisepsis is the reduction of pathogenic microorganisms and viruses on living tissue. Treatment is by chemical antimicrobials, like iodine and alcohol. Antisepsis is used to disinfect living tissues without harming them.

Commercial sterilization is the treatment to kill endospores in commercially canned products. An example is the bacteria *Clostridium botulinum,* which causes botulism.

Aseptic means to be free of pathogenic contaminants. Examples include proper hand washing, flame sterilization of equipment, and preparing surgical environments and instruments.

Any word with the suffix -cide or –cidal indicates the death or destruction of an organism. For example, a bactercide kills bacteria. Other examples are fungicides, germicides and virucides. Germicides include ethylene oxide, propylene oxide, and aldehydes. For the same reason, these germicides are also used in preserving specimens in laboratories.

Disinfection is the destruction or killing of microorganisms and viruses on nonliving tissue by the use of chemical or physical agents. Examples of these chemical agents are phenols, alcohols, aldehydes, and surfactants.

Degerming is the removal of microorganisms by mechanical means, such as cleaning the site of an injection. This area of the skin is degermed by using an alcohol wipe or a piece of cotton swab soaked with alcohol. Hand washing also removes microorganisms by chemical means.

Pasteurization, as noted in Chapter 1, uses heat to kill pathogens and reduce the number of food spoilage microorganisms in foods and beverages. Examples are pasteurized milk and juice.

Sanitation is the treatment to remove or lower microbial counts on objects such as eating and drinking utensils to meet public health standards. This is usually accomplished by washing the utensils in high temperatures or scalding water and disinfectant baths. Bacterostatic, fungistatic, and virustastic agents—or any word with the suffix -static or -stasis—indicate the inhibition of a particular type of microorganism. These are unlike bactericides or fungicides that kill or destroy the organism. Germistatic agents include refrigeration, freezing, and some chemicals.

Microbial Death Rates

Microbial death is the term used to describe the permanent loss of a microorganism's ability to reproduce under normal environmental conditions. A technique for the evaluation of an antimicrobial agent is to calculate the *microbial death* rate. When populations of particular organisms are treated with heat or antimicrobial chemicals, they usually die at a constant rate.

The effectiveness of antimicrobial treatments is influenced by the number of microbes that are present. The larger the population, the longer it takes to destroy it. The different variations of certain microorganisms influence death rate because, for example, endospores are difficult to kill.

Environmental influences, such as the presence of blood, saliva, or fecal matter, inhibits the action of chemical antimicrobials. Time of exposure to heat or radiation is also important. Many chemical antimicrobials need longer exposure times to be effective in the death of more resistant microorganisms or endospores.

Action of Antimicrobial Agents

There are two categories that chemical and physical antimicrobial agents fall into: those that affect the cell walls or cytoplasmic membranes of the microorganism and those that affect cellular metabolism and reproduction. As stated in Chapter 4, the cell wall is located outside the microorganism's plasma membrane. The cell's plasma membrane regulates substances that enter and exit the cell during its life. Nutrients enter the cell as waste products exit the cell. Damage to the plasma membrane proteins or phospholipids by physical or chemical agents allows the contents of the cell to leak out. This causes the death of the cell.

Proteins act as regulators in cellular metabolisms, function as enzymes (which are important in all cellular activities), and form structural components in cell membranes and cytoplasm. The function of a protein depends on its three-dimensional shape. The hydrogen and disulfide bonds between the amino acids that make up the protein maintain this shape. Extreme heat, certain chemicals and very high or low pH can easily break some of these hydrogen bonds. This breakage is referred to as the *denaturing* of the protein. The protein's shape is changed, thus affecting the function of the protein and ultimately bringing death to the cell.

Certain chemicals, radiation, and heat can damage nucleic acids. The nuclear acids, DNA and RNA, carry the cell's genetic information. If these are damaged, the cell can no longer replicate or synthesize enzymes, which are important in cell metabolism.

Chemical Agents That Control Microbial Growth

The growth of a microorganism can be controlled through the use of a chemical agent. A chemical agent is a chemical that either inhibits or enhances the growth of a microorganism. Commonly used chemical agents include phenols, phenolics, glutaraldehyde, and formaldehyde.

PHENOLS AND PHENOLICS

Phenols are compounds derived from pheno (carbolic acid) molecules. Phenolics disrupt the plasma by denaturing proteins; they also disrupt the plasma membrane of the cell. As mentioned in Chapter 1, Joseph Lister used phenol in the late 1800s to reduce infection during surgery.

Alcohols are effective against bacterial fungi and viruses. However, they, are not effective against fungal spores or bacterial endospores. Alcohols that are commonly used are isopropanol (rubbing alcohol) and ethanol (the alcohol we drink).

Alcohols denature proteins and disrupt cytoplasmic membranes. Pure alcohol is not as effective as 70 percent because the denaturing of proteins requires water. Alcohols are good to use because they evaporate rapidly. A disadvantage is that they may not contact the microorganisms long enough to be effective. Alcohol is commonly used in swabbing the skin prior to an injection.

Halogens are nonmetallic, highly resistive chemical elements. Halogens are effective against vegetative bacterial cells, fungal cells, fungal spores, protozoan cysts, and many viruses. Halogen-containing antimicrobial agents include iodine, which inhibits protein function. Iodine is used in surgery and by campers to disinfect water. An iodophur is an iodine-containing compound that is longer-lasting than iodine and does not stain the skin. Other halogen agents include:

- *Chlorine (Cl_2).* Used to treat drinking water, swimming pools, and in sewage plants to treat waste water. Chlorine products such as sodium hypochlorite (household bleach) are effective disinfectants.
- *Chlorine dioxide (ClO_2).* A gas that can disinfect large areas.
- *Chloroamines.* Chemicals containing chlorine and ammonia. They are used as skin antiseptics and in water supplies.
- *Bromine.* Used to disinfect hot tubes because it does not evaporate as quickly as chlorine in high temperatures.
- *Oxidizing agents.* Fill microorganisms by oxidizing their enzymes, thus preventing metabolism. Hydrogen peroxide, for example, disinfects and sterilizes inanimate objects, such as food processing and medical equipment, and is also used in water purification.

Arsenic, zinc, mercury, silver, nickel, and copper are called heavy metals due to their high molecular weights. They inhibit microbial growth because they denature enzymes and alter the three-dimensional shapes of proteins that inhibit or eliminate the protein's function. Heavy metals are bacteriostatic and fungistatic agents.

An example is silver nitrate. At one time, hospitals required newborn babies to receive a one percent cream of silver nitrate to their eyes to prevent blindness caused by *Neisseria gonorhoeae*, which could enter the baby's eyes while passing through the birth canal of a mother who was infected. Today, antibiotic ointments that are less irritating are used. Another example is the use of copper in swimming pools, fish tanks, and in reservoirs to control algae growth. Copper interferes with chlorophyll, thus affecting metabolism and energy.

Aldehydes function in microbial growth by denaturing proteins and inactivating nucleic acids. Two types are glutaraldehyde that is a liquid and formaldehyde that is a gas.

GLUTARALDEHYDE AND FORMALDEHYDE

Glutaraldehyde is used in a two percent solution to kill bacteria, fungi, and viruses on medical and dental equipment. Healthcare workers and morticians dissolve gaseous formaldehyde in water, making a 37 percent solution of *formalin*. Formalin is used in disinfecting dialysis machines, surgical equipment, and embalming bodies after death.

Gaseous agents, such as ethylene oxide, propylene oxide, and beta-propiolactone, are used on equipment that cannot be sterilized easily with heat, chemicals, or radiation. Certain items, like pillows, mattresses, dried or powered food, plasticware, sutures, and heart-lung machines, are placed in a closed chamber, then filled with these gases. Gaseous agents denature proteins.

SURFACTANTS

Surfactants are chemicals that act on surfaces by decreasing the tension of water and disrupting cell membranes. Examples are household soaps and detergents.

Quiz

1. What microorganisms can use oxygen when it is present and continue to grow without it?
   (a) Obligate aerobes
   (b) Facultative anaerobes
   (c) Obligate anaerobes
   (d) Free radicals

2. What microorganisms require oxygen?
   (a) Obligate aerobes
   (b) Facultative anaerobes

  (c) Obligate anaerobes
  (d) Free radicals

3. What microorganisms do not use oxygen and can actually be harmed by it?
  (a) Obligate aerobes
  (b) Facultative anaerobes
  (c) Obligate anaerobes
  (d) Free radicals

4. What is not a toxic form of oxygen?
  (a) Singlet oxygen
  (b) Facultative oxygen
  (c) Superoxide free radicals
  (d) Hydroxyl radical

5. What is a sterile culture medium?
  (a) A hydroxyl radical medium
  (b) A medium containing no living organisms
  (c) A chemically defined medium
  (d) A moisture-filled medium

6. What must chemically defined media contain?
  (a) Growth factors
  (b) Complex media
  (c) Peptone complex
  (d) Protein hydrolysis

7. What is agar?
  (a) A nutrient
  (b) Vitamins
  (c) A solidifying agent
  (d) Broth

8. What is an enrichment culture?
  (a) Something that provides growth for all microorganisms
  (b) Something that inhibits growth for all microorganisms
  (c) An infectious culture
  (d) Something that provides growth for a certain microorganism but not for others

9. What is an inoculating loop?
   (a) A tool used to streak a microorganism in a pure culture
   (b) A tool used to place agar in a pure medium
   (c) A tool used to count colonies of microorganisms
   (d) A tool used to view colonies of microorganisms

10. How can preserved bacteria cultures can be revived?
    (a) Oxygenation
    (b) Hydration
    (c) Hydration and liquid nutrient medium
    (d) Oxygenation and liquid nutrient medium

CHAPTER 7

Microbial Genetics

"Who does she look like, mom or dad?" That's probably one of the first few questions everyone asks when hearing about a new arrival in the family. "Does she have Aunt Jane's eyes? Uncle Joe's nose?" "How about Gramps' sandy hair?" These are some of the many noticeable characteristics that can be passed down from family members.

What is really being asked is what genetic traits did the current generation inherit from previous generations. Think of genetic traits as our computer program; it provides us with instructions on how to do everything needed to stay alive. Some instructions are passed along to the next generation while other instructions are not.

If microorganisms could speak, they might also ask the same questions as we do when a new offspring arrives, because microorganisms also pass along genetic traits to new generations of their species. Those traits preprogram new microorganisms on how to identify and process food, how to excrete waste products, and how to reproduce, as well as nearly everything the microorganism needs to know to survive.

In this chapter you'll learn how microorganisms inherit genetic traits from previous generations of microorganisms.

Genetics

Genetics is the branch of science that studies heredity and how traits (expressed characteristics) are passed to new generations of species and between microorganisms. Scientists who study genetics are called *geneticists* and are interested in how traits are expressed within a cell and how traits determine the characteristics of an organism.

Think of a trait as an instruction that tells an organism how do something, such as how to form a toe. Each instruction is contained in a *gene*. As you can imagine, there are thousands of genes (instructions) necessary for an organism to grow and flourish. This is why if a youngster looks and behaves like her mother, family members tend to say she has her mother's genes—that is, she has more genes (instructions) from her mom than from her dad.

Genes are actually made up of segments or sections of deoxyribonuclear acid (DNA), or in the case of a virus, ribonucleic acid (RNA) molecules. These segments are placed in a specific sequence that code for a functional product.

DNA Replication:
Take My Genes, Please!

In 1868, Swiss biologist Friedrich Miescher carried out chemical studies on the nuclei of white blood cells in pus. His studies led to the discovery of DNA. DNA was not linked to hereditary information until 1943 when work performed by Oswald Avery, Colin MacLeod, and Maclyn McCarty at the Rockefeller Institute revealed that DNA contained genetic information. These studies also revealed that genetic information is passed from "parent cells" to "daughter cells," creating a pathway through which genetic information is passed to the next generation of an organism.

Scientists were baffled about how the exchange of DNA occurred. The answer came in 1953 when American geneticist James Watson and English physicist Francis Crick discovered the double-helical structure of DNA at the University of Cambridge in England. Discovery of the double-helical structure was the key that enabled Watson and Crick to learn how DNA is replicated.

In the late 1950's, Mathew Meselson and Franklin Stahl first described the DNA molecule and how DNA replicates in a process called *semiconservative replication.* DNA is replicated by taking one parent double-stranded DNA molecule, unzipping it and building two identical daughter molecules. Bases along the two strands of double-helical DNA complement each other. One strand of the pair acts as a template for the other.

DNA is replication requires complex cellular proteins that direct the sequence of replication. Replication begins when the parent double-stranded DNA molecule unwinds; then the two strands separate. The DNA polymerase enzyme uses a strand as a template to make a new strand of DNA. The DNA polymerase enzyme examines the new DNA and removes bases that do not match and then continues DNA synthesis.

The point at which the double-stranded DNA molecule unzips is called the *replication fork* (Fig. 7-1). The two new strands of DNA each have a base

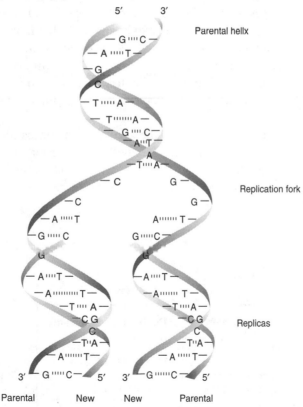

Fig. 7-1.   In semiconservative replication, new strands are synthesized after the replication fork.

sequence complimentary to the original strand. Each double-stranded DNA molecule contains one original and one new strand. In bacteria, each daughter receives a chromosome that is identical to the parent's chromosome.

THE CHROMOSOME CONNECTION

Chromosomes are structures that contain DNA. *DNA* consists of two long chains of repeating nucleotides that twist around each other, forming a double helix. A *nucleotide* in a DNA chain consists of a *nitrogenous base*, a *phosphate group,* and *deoxyribose* (pentose sugar).

The two DNA chains are held together by *hydrogen bonds* between their nitrogenous bases. There are two major types of nitrogenous bases. These are *purines* and *pyrimidines*. There are two types of purine bases: *adenine* (A) and *guanine*(G). There are also two types of pyrimidine bases: *cytosine* (C) and *thymine (T)*. Purine and pyrimidine bases are found in both strands of the double helix.

Base pairs are arrangements of nitrogenous bases according to their hydrogen bonding. *Adenine* pairs with *thymine* and *cytosine* pairs with *guanine*. Adenine is said to be complementary to thymine and cytosine is said to be complementary with guanine. This is known as *complementary base pairing* and is shown in Table 7-1.

Genetic information is encoded by the sequence of bases along a strand of DNA. This information determines how a nucleotide sequence is translated into an amino acid which is the basis of *protein synthesis*. The translation of genetic information from genes to specific proteins occurs in cells.

Table 7-1.   Complementary Messenger RNA
Bases and DNA Bases

Messenger RNA Base	Double-Helical Strand DNA Molecule Base
Adenine (A)	Thymine (T)
Guanine (G)	Cytosine (C)
Cytosine (C)	Guanine (G)
Uricil (U)	Adenine (A)

Protein Synthesis

Protein synthesis is the making of a protein and requires *ribonucleic acid* (RNA), which is synthesized from nucleotides that contain the bases A, C, G, and U (uracil). There are three types of RNA. These are:

- *Ribosomal RNA* (rRNA), which is the enzymatic part of ribosomes.
- *Transfer RNA* (tRNA), which is needed to transport amino acids to the ribosomes in order to synthesize protein.
- *Messenger RNA* (mRNA), which carries the genetic information from DNA into the cytoplasm to ribosomes where the proteins are made.

An enzyme called *RNA polymerase* is required to make (synthesize) rRNA, tRNA, and mRNA.

Protein synthesis begins with the *transcription process,* in which DNA sequences are replicated producing mRNA. The mRNA carries genetic information from the DNA to ribosomes. *Ribosomes* are organelles and the site of protein synthesis.

Nucleotides contained in DNA are duplicated by enzymes before cell division, enabling genetic information to be carried between cells and from one generation to the next. This is referred to as *gene expression* and happens in RNA only.

During transcription, the bases A, C, G, and *uracil* (U) pair with bases of the DNA strand that is being transcribed. The G base in the DNA template pairs with the C base in the mRNA. The C base in the DNA template pairs with the G base in the mRNA. The T base in the DNA template pairs with an A in the mRNA. An A base in the DNA template pairs with the U in the mRNA. This happens because RNA contains a U base instead of a T base.

Transcription begins when the RNA polymerase binds to DNA at the *promotor site.* The DNA unwinds. One of the DNA strands, called a *coding strand,* serves as a template for RNA synthesis. RNA is synthesized by pairing free nucleotides of the RNA with nucleotide bases on the DNA template strand. The RNA polymerase moves along the DNA as the new RNA strand grows. This continues until the RNA polymerase reaches the terminator site on the DNA or is physically stopped by a section of RNA transcript. The new single-stranded mRNA and the RNA polymerase are released from the DNA.

Here is what's happening: Information of the nucleic acid assembles a protein. The mRNA strand consists of several sections, one being the *reading frame.* The reading frame is made up of *codons.* These are AUG, AAA, and GGC. Each codon contains information for a specific amino acid. The sequence of codons on the RNA determines the sequence that amino acids are used to synthesize proteins.

Once the transcription process is completed, information of the mRNA is turned into protein in the translation process. The *translation process* is one in which genetic information encoded in mRNA is translated into a specific sequence of amino acids that produce proteins.

Appropriate amino acids are brought to the translation site in the ribosomes and are assembled into a growing chain. It is here that tRNA recognizes specific codons. Each tRNA molecule has an anticodon, which is a sequence of three bases that is complementary to the bases on the codon. These bases are then paired, and amino acids are brought to the chain.

This process continues until a polypeptide is produced. The polypeptide is removed from the ribosome for further processing. The polypeptide may be stored in the Golgi body of a eukaryotic organism. The mRNA molecule degenerates and the nucleotides are returned to the nucleus. The tRNA molecule is returned to the cytoplasm and combines with new molecules of amino acids.

GENOTYPE AND PHENOTYPE: REALIZING YOUR POTENTIAL

The genetic makeup of an organism is called a *genotype* and represents that organism's potential properties. Some properties may not have developed. Those that do develop are called an organism's phenotype. The *phenotype* represents expressed properties, such as blue eyes and curly hair.

A genotype is the organism's DNA (a collection of genes). The phenotype is a collection of proteins. The majority of the cell's properties comes from the structures and functional properties of these proteins.

Controlling Genes: You're Under My Spell

The process of making proteins (remember, polypeptides become proteins either after they are combined with other polypeptides or when they become biologically functional) begins with the copying of the genetic information found in DNA, into a complimentary strand of RNA. This copying is called *transcription*. Messenger RNA (mRNA) will carry the coded information or instructions for assembling the polypeptides from DNA to the ribosomes of the cell's endoplasmic reticulum where the polypeptides will be made.

The actual building of polypeptides is called *translation*. Translation involves the deciphering of nucleic acid information and converting that information into a language that the proteins can understand.

THE OPERON MODEL

In 1961 Francois Jacob and Jacques Monod formulated the *operon model* that described how transcription of mRNA is regulated. Transcription of mRNA is regulated in two ways. These are repression and induction.

Smoothly, *Repression* inhibits gene expression and decreases the synthesis of enzymes. Proteins called *repressors* stop the ability of RNA polymerase to initiate transcription from repressed genes. *Induction* activates transcription by producing *inducer*, which is the chemical that induces transcription.

Jacob and Monod identified genes in *E. coli* as *structural genes*, *regulatory genes*, and *control regions*. Collectively these form a functional unit called the *operon*. Certain carbohydrates can induce the presence of enzymes needed to digest those carbohydrates.

For example when lactose is present, *E. coli* synthesize enzymes needed to breakdown lactose. Lactose is an inducer molecule. If lactose is absent, a regulator gene produces a repressor protein that binds to a control region called the operator site, preventing the structural genes from encoding the enzyme for lactose digestion. Lactose binds to the repressor at the operator site when lactose is present, freeing the operator site. The structural genes are released and produce their lactose-digesting enzymes.

Mutations: Not a Pretty Copy

A *mutation* is a permanent change in the DNA base sequence (Table 7-2). Some mutations have no expressive effect while other mutations have an expressive effect. When a gene mutates, the enzyme encoded by the gene can become less active or inactive because the sequence of the enzyme amino acids may have changed. The change can be harmful or fatal to the cell, or it can be beneficial— especially if the mutation creates a new metabolic activity.

The most common type of mutation is point mutation, which is also known as *base substitution mutation*. *Point mutation* occurs when an unexpected base is substituted for a normal base, causing alteration of the genetic code, which is then replicated.

Table 7-2. Types of Mutations

Type of Mutation	Description
Point mutation	Also known as *base substitution*, this is the most common type of mutation and involves a single base pair in the DNA molecule. In point mutation, a different base is substituted for the original base, causing the genetic code to be altered. The substituted base pair is used when DNA is replicated.
Missence mutation	A mutation when a new amino acid is substituted in the final protein by the messenger RNA during transcription.
Nonsense mutations	A mutation when a terminator codon in the messenger RNA appears in the middle of a genetic message instead of at the end of the message, which causes premature termination of transcription.
Frame shift mutation	Pairs of nucleotides are either added or removed from a DNA molecule.
Loss-of-function mutation	This mutation causes a gene to malfunction.
Spontaneous mutation	Naturally occurring mutation that happens without the presence of a mutation-causing agent.
Induced mutation	Induced in a laboratory.

If the mutated gene is used for protein synthesis, the mRNA transcribed from the gene carries the incorrect base for that position. The mRNA may insert an incorrect amino acid in the protein. If this happens, the mutation is called a *missence mutation*.

Mutations that change or destroy the genetic code are called *nonsense mutations*. If nucleotides are added or deleted from mRNA, the mutation is called a *frame shift mutation*.

A mutation occurring in the laboratory is called an *induced mutation;* mutations occurring outside the laboratory are called *spontaneous mutation*. A spontaneous mutation occurs when a mutation-causing agent is present.

Base substitution and frame shift mutations occur spontaneously. Agents in the environment or those introduced by industrial processing can directly or indirectly cause mutations. These agents are called mutagens. Any chemical or physical agent that reacts with DNA can potentially cause mutations.

Certain mutations make microorganisms resistant to antibiotics or increase their pathogenicity. There are many naturally occurring mutagens, such as radiation from x-rays, gamma rays, and ultraviolet light. These rays break the covalent bonds between certain bases of DNA-producing fragments.

Ultraviolet light binds together adjacent thymines in a DNA strand, forming *thymine dimers* that cannot function in protein synthesis. Unless repaired, these dimers cause damage or death to cells due to improper transcription or replication of DNA. Some bacteria can repair damage caused by ultraviolet radiation by employing light-repairing enzymes that separate the dimer into the original two thymines. This process is called *photoreactivation*.

MUTATION RATE

Mutations occur naturally and can be induced by mutation-causing agents in the environment. However, not all cells experience mutation even if they are exposed to mutation-causing agents.

Scientists measure the impact that mutation has on an organism by determining the mutation rate. The *mutation rate* is the number of mutations per cell division. For example, suppose you observe the growth of 100 cells that began from a parent cell. If 90 of those cells replicate the parent cell and 10 cells are mutations, than the mutation rate is 10 percent.

Measuring the mutation rate is a way to compare the number of mutations that occur naturally to the number of mutations that occur when a cell is exposed to a potential mutation-causing agent.

First, scientists measure the mutation rate that occurs naturally when a cell is not exposed to a potential mutation-causing agent. Next, the mutation rate is calculated when a cell is not exposed to a potential mutation-causing agent. The results of these two observations are compared. If both mutation rates are relatively the same, then the substance being tested is not a mutation-causing agent. However, the substance is a mutation-causing agent if its mutation rate is appreciably higher than the natural mutation rate.

Quiz

1. The point at which the double-stranded DNA molecule unwinds is called
   (a) polymerase
   (b) replication fork
   (c) hydrogen bonding
   (d) ribosomes

2. Protein synthesis begins with the
   (a) transcription process
   (b) translation process
   (c) polypeptide process
   (d) genotyping

3. Expressed properties such as whether you have blue eyes and curly hair
   is called
   (a) genotype
   (b) exons
   (c) introns
   (d) phenotype

4. Which of the following is not a type of RNA?
   (a) rRNA
   (b) uRNA
   (c) tRNA
   (d) mRNA

5. What is RNA polymerase?
   (a) An enzyme used in the synthesis of RNA
   (b) An enzyme used in the synthesis of ribosomes
   (c) An enzyme used in the synthesis of protein
   (d) An enzyme used in the synthesis of pre-DNA

6. What is the promotor site?
   (a) The site where RNA polymerase binds to DNA
   (b) The site where RNA polymerase binds to protein
   (c) The site where RNA polymerase binds to free nucleotides
   (d) The site where RNA polymerase binds to AAA

7. The mRNA language consists of three nucleotides called
   (a) amino acid
   (b) codons
   (c) introns
   (d) exons

8. What do repressors do?
   (a) They activate transcriptions.
   (b) They increase synthesis.

(c) They increase gene expression.

(d) They stop the initiation of transcription.

9. An operon consists of structural genes, regulatory genes, and control genes.

   (a) True

   (b) False

10. Spontaneous mutation occurs as a result of laboratory intervention in DNA replication.

   (a) True

   (b) False

CHAPTER 8

Recombinant DNA Technology

Our genes are a strong determining factor of who we are and what we are going to be because genes program our bodies to express specific characteristics. Some of those characteristics enable us to carry out basic life functions, such as converting food to energy, while others make us stand out in a crowd, such as being a seven-foot professional basketball player. The same concept holds true with microorganisms. A microorganism's genes determine characteristics expressed by that microorganism.

Genetic information is encoded in our DNA by the linkage of nucleic acids in a specific sequence, which you learned about in the previous chapter. Think of what would happen if you could change the sequence. You could reprogram genes to express desirable characteristics and to repress undesirable characteristics, such as those that cause diseases.

Reordering genetic information is called genetic engineering. In this chapter, you'll learn about genetic engineering and how to use recombinant DNA technology to alter the genetic program of an organism.

Genetic Engineering: Designer Genes

The modification of an organism's genetic information by changing its nucleic acid genome is called *genetic engineering* and is accomplished by methods known as *recombinant DNA technology*. Recombinant DNA technology opens up totally new areas in research and applied biology and is an important part of biotechnology, a field that is increasingly growing. *Biotechnology* is the term used for processes in which organisms are manipulated at the genetic level to form products for medicine, agriculture, and industry.

Recombinant DNA is DNA with a new sequence formed by joining fragments from different sources. One of the first breakthroughs leading to recombinant DNA, or rDNA, technology was the discovery of microbial enzymes that make cuts into the double-stranded DNA. These were discovered by Werner Arber, Hamilton Smith, and Dan Nathans in the late 1960s. These enzymes recognize and cleave specific sequences of four to eight base pairs and are known as *restriction enzymes*. These enzymes recognize specific sequences in DNA and then cut the DNA to produce fragments called *restriction fragments*. The enzymes cut the bonds of the DNA backbone at a point along the exterior of the DNA strands.

There are three types of restriction enzymes. Types I and III cleave DNA away from recognition sites. Type II restriction endonucleases cleave DNA at specific recognition sites. The type II enzymes can be used to prepare DNA fragments containing specific genes or portions of genes. A gene can be defined as a segment of DNA (a segment is a sequence of nucleotides) that codes for a functional product.

ECORI cleaves the DNA between guanine (G) and adenine (A) in the base sequence GAATTC. In the double-stranded condition, the base sequence GAATTC will base pair with a sequence, which runs in the opposite direction. ECORI cleaves both DNA strands between the G and the A. When the two DNA fragments separate they contain single-stranded complementary ends called sticky ends.

Each restriction enzyme name begins with the first three letters of the bacterium that produces it. This is illustrated in Table 8-1.

In 1972, David Jackson, Robert Symons, and Paul Berg generated recombinant DNA molecules. They allowed the sticky ends of the fragments to base pair with each other and covalently joined the fragments with the enzyme DNA ligase. The enzyme DNA ligase links the two sticky ends of the DNA molecules at the point of union. In 1973, Stanley Cohen and Herbert Boyer constructed the first recombinant plasmid capable of being replicated within a bacterial host. A *plasmid* is a circular DNA molecule that a bacterium can replicate without a chromosome.

Table 8-1. Recombinant DNA Is DNA with a New Sequence Formed by Joining Recognition Sequence Fragments from Different Sources

Enzyme	Microbial Source	Recognition Sequence	Cleavage Sites ($\downarrow$,$\uparrow$)	End Product	
EcoRI	*Escherichia coli*	GAATTC	G$\downarrow$AATTC	G	AATTC
		CTTAAG	CTTAA$\uparrow$G	CTTAA	G

In 1975, Edwin M. Southern developed procedures for detecting specific DNA fragments so that a particular gene could be isolated from a complex DNA mixture. This technique is called the *Southern blotting technique*. DNA fragments are separated by size with *agarose gel electrophoresis*. Gel electrophoresis takes advantage of the chemical and physical properties of DNA to separate the fragments. The phosphate groups in the backbone of DNA are negatively charged. This makes the DNA molecules attracted to anything that is positively charged. In gel electrophoresis the DNA molecules are placed in an electric field so that they migrate towards the positive charge.

The DNA is placed in agarose, a semi solid gelatin, and placed in a tank of buffer. When electrical current is applied, the DNA molecules migrate through the agarose gel, separate, and travel toward the positive poles of the electric fields. The entire DNA fragments migrates through the gel. The larger DNA fragments have a harder time moving than the smaller ones, so the small fragments travel farther through the gel.

ARTIFICIAL DNA: PUTTING TOGETHER THE PIECES

Oligonucleotides, from the Greek word *oligo* meaning "few," are short pieces of DNA or RNA that are 2 to 30 nucleotides long. The ability to synthesize DNA oligonucleotides of a known sequence is incredibly important and useful. A DNA probe is used to analyze fragments of DNA. A DNA probe is a single-stranded fragment of DNA that recognizes and binds to a complementary section of DNA in a mixture of DNA molecules.

DNA probes can be synthesized and DNA fragments can be prepared for use in molecular techniques such as *polymerase chain reaction* (PCR). Polymerase chain reaction is a technique that was developed by Kary Mullis in 1985. It produces large quantities of a DNA fragment without needing a living cell. Starting with one small piece of DNA, PCR can make billions of copies in a few hours. These large quantities of DNA can be easily analyzed.

PCR and DNA probes have been of great value to the areas of molecular biology, medicine, and biotechnology. Using these tools, scientists can detect the DNA associated with HIV (the virus that causes AIDS), Lyme disease, chlamydia, tuberculosis, hepatitis, HPV (human papilloma virus), cystic fibrosis, muscular distrophy, and Huntington's disease.

Gene Therapy: Makes You Feel Better

Gene therapy is a recombinant DNA process, in which cells are taken from the patient, altered by adding genes, and returned to the patient. A type of genetic surgery called *somatic gene therapy* may be possible. Cells of a person with a genetic disease could be removed, cultured, and transformed with cloned DNA containing a normal copy of the defective gene. They could be reintroduced into the individual. If these cells become established, the expression of the normal genes may be able to cure the patient.

In the early 1990s, gene therapy of this type was used to correct a deficiency of the enzyme *adenosine deaminase* (ADA). An immune deficiency disease patient lacking the enzyme adenosine deaminase, an enzyme that destroys toxic metabolic byproducts, had been treated. Some of the patient's lymphocytes were removed. Lymphocytes are a type of white blood cell that fights infection. The lymphocytes were given the adenosine deaminase gene with the use of a modified retrovirus—which served as a *vector*—and placed back into the patient's body. Once established in the body, the cells with altered genes began to make the enzyme adenosine deaminase (ADA) and alleviated the deficiency.

DNA Fingerprinting: Gotcha

DNA fingerprinting is an area of molecular biology that involves analyzing genetic material. It involves the use of restriction enzymes, which cut DNA molecules into pieces. When DNA samples obtained from different individuals are cut with the same restriction enzyme, the number and size of restriction fragments produced may be different. This difference provides the basis for DNA fingerprinting. The use of DNA fingerprinting depends upon the presence of repeating base sequences. These sequences are called *restriction fragment length polymorphosis*, or the RFLP pattern, which is unique for every individual.

This is a sort of molecular signature or fingerprint. In order to perform DNA fingerprinting, DNA must be taken from an individual. Samples can be taken from hair, blood, skin, cheek cells, or other tissue. The DNA is taken from the cells and is broken down with enzymes. The fragments are separated with *electrophoresis*. The DNA fragments are then analyzed for RFLPs using DNA probes. An evaluation enables crime lab scientists (forensic pathologists) to compare a person's DNA with the DNA taken from a scene of a crime. This technique has a 99 percent degree of certainty that a suspect was at a crime scene.

INDUSTRIAL APPLICATION: SHOW ME THE MONEY

Industrial applications of recombinant DNA technology include manufacturing protein products by the use of bacteria, fungi, and cultured mammal cells. The pharmaceutical industry is producing several medically important polypeptides using biotechnology. An example is bacteria that metabolize petroleum and other toxic materials. These bacteria are constructed by assembling catabolic genes on a single plasmid and then transforming the appropriate organism. Another example is *vaccines*. The hepatitis B vaccine is made up of viral protein manufactured by yeast cells, which have been *recombined* with viral particles.

AGRICULTURAL APPLICATIONS: CROPS AND COWS

Recombinant DNA and biotechnology have been used to increase plant growth by increasing the efficiency of the plant's ability to fix nitrogen. Scientists take genes for nitrogen fixation from bacteria and place the genes into plant cells. Because of this, plants can obtain nitrogen directly from the atmosphere. The plants can produce their own proteins without the need for bacteria. Another way to insert genes into plants is with a recombinant *tumor-inducing plasmid Ti plasmid*. This is obtained from the bacterium *Agrobacterium tumefaciens*. This bacteria invades plant cells and its plasmids insert chromosomes that carry the genes for tumor induction. An example of recombinant DNA with livestock is the recombinant bovine growth hormone that has been used to increase milk production in cows by 10 percent.

U.S. farmers grow substantial amounts of genetically modified crops. About one-third of the corn and one-half of the soybean and cotton crops are genetically modified. Cotton and corn have become resistant to herbicides and insects. Soybeans have herbicide resistance and lower saturated fat content. Having herbicidal-resistant plants is important because many crop plants suffer stress when treated with herbicides. Resistant crops are not stressed by the chemicals that are used to control weeds.

Recombinant DNA Technology and Society: Too Much of a Good Thing

Genetically altering an organism raises scientific and philosophical questions. Recombinant DNA technology has had a positive impact on society, although there may be associated dangers with rDNA.

There have been concerns raised by the scientific community that genetically engineered microorganisms carrying dangerous genes might be released into the environment and cause widespread infection. Due to these worries, the federal government has established guidelines to regulate and limit the locations and types of experiments that are potentially dangerous.

Biomedical rDNA research has been regulated by the Recombinant DNA Advisory Committee (RAC) of the Natural Institutes of Health. The Food and Drug Administration (FDA) has principal responsibility in overseeing gene therapy research. The Environmental Protection Agency (EPA) and state governments have jurisdiction over field experiments in agriculture.

One of the biggest efforts in biotechnology has been the *human genome project*, which began in 1990 and formally ended in 2001. The goal of this project has been to determine the sequences of all human chromosomes. Advances like this in biotechnology will make genetic screening incredibly effective. Physicians will one day be capable of detecting genetic flaws in DNA long before the disease becomes manifested in a patient.

Another area of controversy is agriculture. Some scientists state that the release of recombinant organisms without risk assessment may disrupt the ecosystem. Viral nucleic acids, inserted into plants to make them resistant to viruses, might combine with the genome of an invading virus to make the virus even stronger. Genetically modified food might even trigger an allergic response in people or animals that consume them. As of this writing, obvious health or ecological events have not been observed. However, due to the consensus of the public, many food producers have stopped using genetically modified crops.

Quiz

1. The nucleotide genome consists of the sequence of nucleic acid that encodes genetic information on DNA.
   (a) True
   (b) False

2. What enables scientists to take nucleotide fragments from other DNA and reassemble fragments into a new nucleotide sequence?
   (a) Enzyme DNA technology
   (b) Enzyme technology
   (c) Recombinant DNA technology
   (d) Recombinant enzyme technology

3. What is used to cut DNA double-helix strand DNA along the exterior of the strand?
   (a) Overhang
   (b) Restriction enzymes
   (c) Restriction fragment
   (d) Recognition sequence

4. What is the particular nucleotide sequence of a double-helical segment called?
   (a) Overhang
   (b) Restriction enzymes
   (c) Restriction fragment
   (d) Recognition sequence

5. What is the end of the cut of a double-helical segment called?
   (a) Overhang
   (b) Restriction enzymes
   (c) Restriction fragment
   (d) Recognition sequence

6. What results when two incisions are made in a double-helical segment?
   (a) Overhang
   (b) Restriction enzymes
   (c) Restriction fragment
   (d) Recognition sequence

7. The four nucleotides are adenine (A), cytosine (C), guanine (G) and thymine (T).
   (a) True
   (b) False

8. What is another name for a restriction enzyme?
   (a) Vector
   (b) Plasmid
   (c) Restriction endonucleases
   (d) Agarose gel

9. The Southern blotting technique is used for detecting specific restriction fragments.
   (a) True
   (b) False

10. Scientists synthesize fragments of DNA and RNA using a process known as polymerase chain reaction (PCR).
    (a) True
    (b) False

CHAPTER 9

Classification of Microorganisms

It is uncanny how a stranger can be at a large family gathering and be able to pick out who belongs to Mom's family and who belongs to Dad's family. Although no two people at the gathering look the same, there are enough similarities among some for the unbiased observer to deduce a relationship.

Scientists group together microorganisms much the same way as a stranger can group family members together. That is, scientists carefully observe microorganisms and classify them into groups based on similar characteristics.

In this chapter, you'll learn how to use scientific techniques to organize microorganisms into standard classifications based on a microorganism's characteristics.

Taxonomy: Nothing to Do with the IRS

Organisms have traits that are similar to and different from other organisms. Scientists organize organisms into groups by developing a taxonomy. *Taxonomy* is the science of organisms based on a presumed natural relationship.

Scientists observe each organism, noting its characteristics. Organisms that have similar characteristics are presumed to have a natural relationship and therefore are placed in the same group. Classification tries to show this natural relationship.

Taxonomy has three components:

- *Classification.* The arrangement of organisms into groups based on similar characteristics, evolutionary similarity or common ancestry. These groups are also called *taxa.*

- *Nomenclature.* The name given to each organism. Each name must be unique and should depict the dominant characteristic of the organism.

- *Identification.* The process of observing and classifying organisms into a standard group that is recognized throughout the biological community.

Taxonomy is a subset of systemics. *Systemics* is the study of organisms in order to place organisms having similar characteristics into the same group. Using techniques from other sciences such as biochemistry, ecology, epidemiology, molecular biology, morphology, and physiology, biologists are able to identify characteristics of a organism.

BENEFITS OF TAXONOMY

Taxonomy organizes large amounts of information about organisms whose members of a particular group share many characteristics. Taxonomy lets scientists make predictions and design a hypothesis for future research on the knowledge of similar organisms. A hypothesis is a possible explanation for an observation that needs experimentation and testing.

If a relative of an organism has the same properties, the organism may also have the same characteristics. Taxonomy puts microorganisms into groups with precise names, enabling microbiologists to communicate with each other in an efficient manner. Taxonomy is indispensable for the accurate identification of microorganisms. For example, once a microbiologist or epidemiologist identifies a pathogen that infects a patient, physicians know the proper treatment that will cure the patient.

Nomenclature of Taxonomy: Name Calling

In the mid-1700s, Swedish botanist Carl Linnaeus was one of the first scientists to develop a taxonomy for living organisms. It is for this reason that he is known as the father of taxonomy. Linnaeus' taxonomy grouped living things into two kingdoms: plants and animals.

By the 1900s, scientists had discovered microorganisms that had characteristics that were dramatically different than those of plants and animals. Therefore, Linnaeus' taxonomy needed to be enhanced to encompass microorganisms.

In 1969 Robert H. Whitteker, working at Cornell University, proposed a new taxonomy that consisted of five kingdoms (see Fig. 9-1). These were monera, protista, plantae (plants), fungi, and animalia (animals). *Monera* are organisms that lack a nucleus and membrane-bounded organelles, such as bacteria. *Protista* are organisms that have either a single cell or no distinct tissues and organs, such as protozoa. This group includes unicellular eukaryotes and algae. *Fungi* are organisms that use absorption to acquire food. These include multicellular fungi and single-cell yeast. Animalia and plantae include only multicellular organisms.

Scientists widely accepted Whitteker's taxonomy until 1977 when Carl Woese, in collaboration with Ralph S. Wolfe at the University of Illinois, proposed a

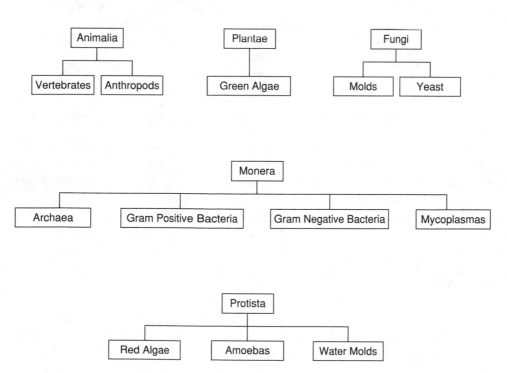

Fig. 9-1.   Whitteker's five-kingdom taxonomy.

new six-kingdom taxonomy. This came about with the discovery of archaea, which are prokaryotes that lives in oxygen-deprived environments.

Before Woese's six-kingdom taxonomy, scientists grouped organisms into eukaryotes animals, plants, fungi, and one-cell microorganisms (paramecia)— and prokaryotes (microscopic organisms that are not eukaryotes).

Woese's six-kingdom taxonomy consists of:

- Eubacteria (has rigid cell wall)
- Archaebacteria (anaerobes that live in swamps, marshes, and in the intestines of mammals)
- Protista (unicellular eukaryotes and algae)
- Fungi (multicellular forms and single-cell yeasts)
- Plantae
- Animalia

Woese determined that archaebacteria and eubacteria are two groups by studying the rRNA sequences in prokaryotic cells.

Woese used three major criteria to define his six kingdoms. These are:

- *Cell type.* Eukaryotic cells (cells having a distinct nucleus) and prokaryotic cell (cells not having a distinct nucleus)
- *Level of organization.* Organisms that live in a colony or alone and one-cell organisms and multicell organisms.
- *Nutrition.* Ingestion (animal), absorption (fungi), or photosynthesis (plants).

In the 1990s Woese studied rRNA sequences in prokaryotic cells (archaebacteria and eubacteria) proving that these organisms should be divided into two distinct groups. Today organisms are grouped into three categories called domains that are represented as bacteria, archaea, and eukaryotes.

The domains are placed above the phylum and kingdom levels. The term archaebacteria (meaning from the Greek word *archaio* "ancient") refers to the ancient origin of this group of bacteria that appears to have diverged from eubacteria. The archaea and bacteria came from the development of eukaryotic organisms.

The evolutionary relationship among the three domains is:

- Domain Bacteria (eubacteria)
- Domain Archaea (archaebacteria)
- Domain Eulcarya (eukaryotes)

Different classifications of organisms are:

- Bacteria
  - Eubacteria
- Archaea
  - Archaebacteria
- Eukarya
  - Protista
  - Fungi
  - Plantae
  - Animalia

The three domains are archaea, bacteria, and eukarya (see Fig. 9-2).

- *Archaea* lack muramic acid in the cell walls.

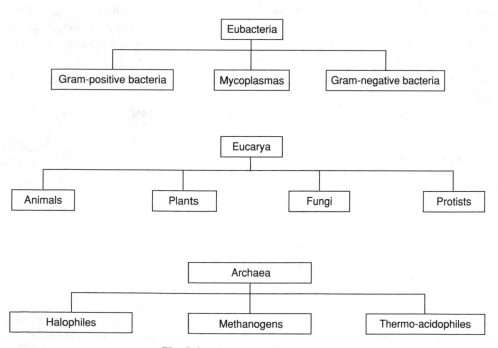

Fig. 9-2. Three-domain taxonomy.

- *Bacteria* have a cell wall composed of peptidoglycan and muramic acid. Bacteria also have membrane lipids with ester-linked, straight-chained fatty acids that resemble eukaryotic membrane lipids. Most prokaryotes are bacteria. Bacteria also have plasmids, which are small, double-stranded DNA molecules that are extrachromosomal.

- Eukarya are of the domain eukarya and have a defined nucleus and membrane bound organelles.

TAXONOMIC RANK AND FILE

A taxonomy has an overlapping hierarchy that forms levels of *rank* or *category* similar to an organization chart. Each rank contains microorganisms that have similar characteristics. A rank can also have other ranks that contain microorganisms.

Microorganisms that belong to a lower rank have characteristics that are associated with a higher rank to which the lower rank belongs. However, characteristics of microorganisms of a lower rank are not found in microorganisms that belong to the same higher rank as the lower-rank microorganism.

Microbiologists use a *microbial taxonomy* (Fig. 9-3), which is different from what biologists, who work with larger organisms, use. Microbial taxonomy is commonly called *prokaryotic taxonomy*. The widely accepted prokaryotic taxonomy is *Bergey's Manual of Systematic Bacteriology,* first published in 1923 by the American Society for Microbiology. David Bergey was chairperson of the editorial board.

In the taxonomy of prokaryotes, the most commonly used rank (in order from most general to most specific) is:

Domain
　Kingdom
　　Phyla
　　　Class
　　　　Order
　　　　　Family
　　　　　　Genus
　　　　　　　Species

The basic taxonomic group in microbial taxonomy is the species. Taxonomists working with higher organisms define their species differently than microbiologists. Prokaryotic species are characterized by differences in their phenotype and

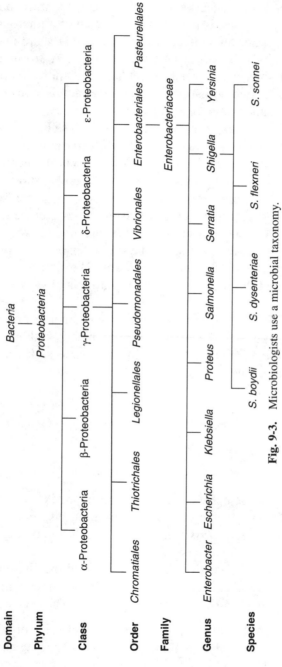

Fig. 9-3. Microbiologists use a microbial taxonomy.

genotype. *Phenotype* is the collection of visible characteristics and the behavior of a microorganism. *Genotype* is the genetic make up of a microorganism.

The prokaryotic species are collections of strains that share many properties and differ dramatically from other groups or strains. A *strain* is a group of microorganisms that share characteristics that are different from microorganisms in other strains. Each microorganism within a strain is considered to have descended from the same microorganism.

For example, *Biovars* is a species that contains strains characterized by differences in its biochemistry and physiology. *Morphovars* is also a species whose strains differ morphologically and structurally. *Serovars* is another species that has strains that are characterized by distinct antigenic properties (substances that stimulate the production of antibodies).

Microbiologists use the genus of the taxonomy to name microorganisms, which you learned in Chapter 1. Microorganisms are given a two-part name. The first part is the Latin name for the genus. The second part is the epithet. Together these parts uniquely identify the microorganism. The first part of the name is always capitalized and the second part of the name is always lowercase. Both parts are italicized.

For example, *Escherichia coli* is a bacterium that is a member of the Escherichia genus and has the epithet *coli*. Sometimes the name is abbreviated such as *E. coli*. However, the abbreviation maintains the same style as the full name (uppercase, lowercase, italic).

Classification: All Natural

A taxonomy is based on scientists' ability to characterize organisms into a classification system. The most widely used classification system is called the natural classification. The *natural classification* requires that an organism be grouped with organisms that have the same characteristics.

In the mid-eighteenth century, Linnaeus developed the first natural classification using anatomical characteristics of organisms. Other natural classifications use classical characteristics to group organisms. These characteristics are:

- *Morphological.* Morphological characteristics classify organisms by their structure, which normally remain the same in a changing environment and are good indications of phylogentic relatedness.

- *Ecological.* Ecological characteristics classify organisms by the environment in which they live. For example, some microorganisms live in various parts of the human intestines and others live in marine environments. Ecological characteristics include the ability to cause disease, temperature, pH, and oxygen requirements of an organisms, as well as an organism's life cycle.

- *Genetic.* Genetic characteristics classify organisms by the way in which they reproduce and exchange chromosomes. For example, eukaryotic organisms reproduce sexually by conjugation where two cells come together and exchange genetic material. Prokaryotic organisms do not reproduce sexually and instead use transformation to reproduce. Transformation occurs between strains of prokaryotes if their genomes are dissimilar but rarely between genera.

In the early 1990s, T. Cavalier-Smith developed the two-empire and eight-kingdom taxonomy based on phentic and phylogenetic characteristics. Phentic measures the physical characteristics of an organism using a process called numerical taxonomy. *Numerical taxonomy* is a phentic classification based on physical measurements of an organism. Phylogenetic measures the evolutionary relationship among organisms.

The two empires are bacteria and eukaryota. The bacteria domain contains two kingdoms. These are eubacteria and archaeobacteria. The eukaryota empire contains six kingdoms as shown in Table 9-1.

Table 9-1.   Cavalier-Smith Two-Empire and Eight-Kingdom Taxonomy

Empire	Kingdom
Bacteria	Eubacteria—Large group of bacteria that have rigid cell walls.
	Archaeobacteria—nonrigid cell walls.
Eukaryota	Archezoa—Primitive one-cell eukaryotes.
	Chromista—Photosynthetic organisms that have chloroplasts within the lumen of the rough endoplasmic reticulum.
	Plantae—Photosynthetic organisms that have chloroplasts in the cytoplasmic matrix.
	Fungi—Absorb nutrients.
	Animalia—Ingest nutrients.
	Protozoa—Single-cell organism.

Quiz

1. Taxonomy is the classification of organisms based on a presumed natural relationship.
   (a) True
   (b) False

2. The arrangement of organisms into groups based on similar characteristics is called
   (a) nomenclature
   (b) identification
   (c) classification
   (d) systemics

3. The name given to each group of organism is called
   (a) nomenclature
   (b) identification
   (c) classification
   (d) systemics

4. The process of observing and classifying organisms into a standard group is called
   (a) nomenclature
   (b) identification
   (c) classification
   (d) systemics

5. The study of organisms in order to place organisms into groups is called
   (a) nomenclature
   (b) identification
   (c) classification
   (d) systemics

6. An animal is an organism that ingests food.
   (a) True
   (b) False

7. What acquires nutrients through absorption?
   (a) Animals
   (b) Plants

    (c) Fungi
    (d) Humans

8. What acquires nutrients through photosynthesis?
    (a) Animals
    (b) Plants
    (c) Fungi
    (d) Humans

9. Bacteria have a cell wall composed of peptidoglycan and muramic acid.
    (a) True
    (b) False

10. A *genus* consists of one or more lower ranks called species.
    (a) True
    (b) False

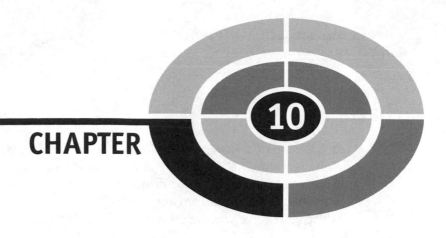

CHAPTER 10

The Prokaryotes: Domains Archaea and Bacteria

Mention the term bacteria and you probably think back to a time when you had a bacterial infection. Some bacteria do cause disease, but other bacteria are beneficial and live within our bodies, aiding in the digestion of food.

There are many different kinds of bacteria, all of which are prokaryotes.

Bacteria are divided into four divisions called *phyla* based on characteristics of their cell wall. Each division is subdivided into sections according to other characteristics, such as oxygen requirements, motility, shape, and Gram-stain reaction. Each section is named based on these characteristics. Sections are further subdivided into genera.

In this chapter, you'll learn about major types of bacteria and their characteristics.

Archaea

Archaea can exist in very hot and very cold environments, making them resilient microorganisms that can survive attacks that destroy other bacteria. For example, archaea can survive and grow in an oxygen-free environment (anaerobic) and in a high-salt (hypersaline) environment.

There are three ways microbiologists identify archaea. Archaea:

- Have a unique sequence of rRNA.
- Have cell walls that lack peptidoglycan. The cell wall of most bacteria contains peptidoglycan.
- Have a membrane lipid that has a branched chain of hydrocarbons connected to glycerol ester links. The membrane lipid of most bacteria has glycerol connected to fatty acids by ester bonds.

Unfortunately, two of the more common techniques used to identify bacteria are not very useful in identifying archaea. You'll recall from Chapter 4 that microbiologists identify bacteria by using the Gram stain. A bacterium is either gram-positive or gram-negative. However, archaea could be gram-positive or gram-negative, which makes the Gram stain test useless when trying to identify archaea.

The shape of a bacterium is another common way microbiologists identify bacteria. Many bacteria have a distinctive appearance. However, archaea are *pleomorphic*, which means they can have various shapes. Sometimes archaea are spiracle, spiral, lobed, plate-shaped, or irregularly shaped.

Archaea also have various types of metabolism. Some archaea are organotrophs while others are autotrophs. Archaea also break down (catabolize) glucose for energy in various ways. It is these variations that enable archaea to survive in environments that are fatal to other bacteria.

THE ARCHAEA CLAN

Archaea are not bacteria and can be organized into subgroups. Microbiologists use one of two subgroup classifications for archaea. One classification method divides archaea into five subgroups. These are:

- *Methanogenic archaea.* A single-celled archaea that produces methane and carbon dioxide (CO_2) through the fermentation of simple organic carbon compounds or the oxidation of H_2 without oxygen to produce CO_2.

- *Sulfate reducers.* Archaea that function in the presence of air.
- *Extreme halophiles.* Archaea that live in an extremely salty environment.
- *Cell wall–less archaea.* Archaea that do not have a cell wall.
- *Extremely thermophilic S-metabolizers.* Archaea that need sulfur for growth.

The other method used to organize archaea into subgroups is used in *Bergey's Manual of Systematic Bacteriology* that you learned about in Chapter 9; it consists of two branches (phyla). These are:

- *Phylum crenarchaeota.* Archaea that are within the phylum crenarchaeota branch are anaerobes (they live in the absence of oxygen) and grow in a sulfur-enriched soil or water environment that is at a temperature between 88 and 100 degrees Fahrenheit and has a pH between 0 and 5.5. Extremely thermophilic S-metabolizers are within the phylum crenarchaeota subgroup.

- *Phylum euryarchaeota.* The phylum euryarchaeota branch consists of the following five major groups:
  - *Methogenic archaea.* Methogenic archaea, the largest group of phylum euryarchaeota, are anaerobic archaea that synthesize organic compounds in a process called *methanogenesis*, which produces methane. They also use inorganic sources (autotrophic) such as H_2 and CO_2 for growth. Methogenic archaea thrive in swamps, hot springs, and fresh water as well as in marshes. They digest sludge and transform undigested food, in animal intestines and in the rumen of a ruminant, into methane. A *ruminant* is a herbivoir that has a stomachs which is divided into four compartments. The *rumen* is the expanded upper compartment of the stomach that contains regurgitated and partially digested food called a cud. Methogenic archaea transform regurgitated and partially digested food into methane (CH_4), which is a clean-burning fuel. For example, a cow can belch up to 400 liters of methane a day. Sewage treatment plants also use methogenic archaea to transform organic waste into methane. Although methane is a source of energy, it is also a cause for the greenhouse effect. Methogenic archaea are further organized into five orders. These are: methanobacterioles, methanococcales, methanomicrobiales, methanosareinales, and methanopyrales.
  - *Extreme halophiles.* Extreme halophiles, also known as *halobacteria*, absorb nutrients from dead organic matter absorb nutrients in the presents of oxygen (*aerobic chemoheterotrophs*). They require proteins, amino acids, and other nutrients for growth in a high concentration of sodium chloride. Extreme halophiles can be motile or nonmotile and are found

in salt lakes and in salted fish and turn lakes and fish red when there is an abundance of Extreme halophiles.

- *Halobacterium salinarium.* Halobacterium salinarium is an archaea that acquires energy through photosynthesis. However, it is able to do so without the need of chlorophyll or bacteriochlorophyll. Halobacterium salinarium synthesizes the *bacteriorhodopsin* protein,which shows as a deep purple color under high-intensity lighting in a low-oxygen environment.

- *Thermophilic archaeon.* Thermophilic archaeons are known as *thermoplasma* and grow in hot (55 to 59 degrees Celsius), acidic (pH of 1 to 2) refuse piles of coalmines that contain iron pyrite. These refuse piles become hot and acidic as chemolithotropic bacteria oxidize iron pyrite into sulfuric acid. Thermophilic archaeons lack a cell wall.

- *Sulfate-reducing archaea.* Sulfate-reducing archaea are known as *archaeoglobi* and extract electrons from various donors to reduce sulfur to sulfide in an environment that is approximately 83 degrees Celsius such as near *marine hydrothermal vents* (underwater hot springs). Sulfate-reducing archaea are gram-negative and are shaped as irregular spheres (coccoid cells).

Aerobic/Microaerophilic, Motile, Helical/Vibroid, Gram-Negative Bacteria

Another kind of prokaryote is the aerobic/microaerophilic, motile, helical/vibroid, gram-negative bacterium. This is a mouthful to say, but the name describes characteristics of this group of prokaryote bacteria.

Aerobic/microaerophilic means bacteria within this group require small amounts of oxygen to grow. *Motile* implies that the bacterium is self-propelled, using flagella at one or both poles to move in a corkscrew motion. *Helical/vibroid* indicates that the bacterium takes the shape of a spiral (helical) or as a curved rod (vibroid). Gram-negative means that when the bacterium is identified using the Gram stain, the bacterium loses the violet stain when rinsed and appears red or pink.

Aerobic/microaerophilic, motile, helical/vibroid, gram-negative bacteria thrive in soil and are found on roots of plants such as the *Azospirillum*, which improves a plant's nutrient uptake. Bacteria within this group are also found in both fresh and stagnant water.

Some aerobic/microaerophilic, motile, helical/vibroid, gram-negative bacteria cause diseases (pathogenic) such as *Campylobacter fetus* and *Campylobacter jejuni. Campylobacter fetus* causes spontaneous abortion in domestic animals. *Campylobacter jejuni* causes inflammation of the digestive tract (enteritis) result-ing in food-borne intestinal diseases. Another common aerobic/microaerophilic, motile, helical/vibroid, gram-negative bacterium that is pathogenic is *Helicobacter pylori. Helicobacter pylori* cause gastric ulcers in humans.

Gram-Negative Aerobic Rods and Cocci

Bacteria that are members of the gram-negative aerobic rods and cocci group include many bacteria that cause disease in humans and bacteria that are impor-tant to industry and the environment. There are 11 bacteria in this group:

- *Pseudomonads. Pseudomonads* are rod-shaped bacteria with polar flagella, which give the bacteria mobility. They need oxygen to grow and obtain energy by decomposing organic material. Pseudomonads are found in soil, fresh water and marine environments.

- *Pseudomonas aeruginosa. Pseudomonas aeruginosa* is a pathogenic bac-terium that infects the urinary tract and wounds in humans. It also causes infections in burn injuries.

- *Legionella pneumophilia. Legionella pneumophilia* is a bacterium identi-fied in 1976 when it infected and killed members of the American Legion at their convention in Philadelphia. Infection caused by the *Legionella pneumophilia* bacterium is commonly referred to as Legionnaire's disease.

- *Legionella micdadei. Legionella micdadei* is the bacterium that infects lungs and causes a strain of pneumonia commonly called Pittsburgh pneumonia.

- *Moraxella lacunata. Moraxella lacunata* is an egg-shaped (coccobacilli) bacterium that can infect the membrane that lines eyelids called the con-juntiva, causing a condition known as conjunctivitis (pink eye).

- *Neisseria. Neisseria* is a double-spherical (diplo-coccus) bacterium that can live with or without oxygen (anaerobic) and is usually found on mucous membranes of humans. One type of *Neisseria* called *Neisseria gonorrhoeae* causes the sexually transmitted disease gonorrhea. Another type is *Neisseria meningitidis* (*N. meningitis*), the bacterium that infects the mucous membranes of the nose and throat (*nasopharyngeal*), causing a sore throat. *Neisseria meningitidis* can cause meningitis if the bacterium enters blood and cerebral spinal fluid where it can infect the protective covering (meninges) of the brain and spinal cord. Meningitis is inflamma-tion of the meninges.

- *Brucella. Brucella* bacteria are very small coccobacillus that can not move themselves (non-motile) and cause brucellosis or undulant fever—daily episodes of fever and chills. *Brucella* multiply in white blood cells.

- *Bordetella pertussis. Bordetella pertussis* is the bacteria that cause pertussis, which is better known as whooping cough. *Bordetella pertussis* is rod-shaped and non-motile.

- *Franeisella tularensis. Franeisella tularensis* is a gram-negative coccobacillus bacterium that lives in contaminated water or wild game. When such water or wild game is ingested, the bacteria infects the lymph nodes (lymphadenopathy), causing a disease called tularemia, which is commonly known as rabbit fever or deer-fly fever. *Franeisella tularensis* can also be inhaled during the skinning of an infected animal or enter through a lesion in the body.

- *Agrobacterium tumefaciens. Agrobacterium tumefaciens* is a bacterium that causes tumor-like growths on plants called crowngull.

- *Acetobacter* and *gluconobacter. Acetobacter* and *gluconobacter* are bacteria that synthesize ethanol to vinegar (acetic acid) and are used in the food industry to make vinegar.

Facultatively Anaerobic Gram-Negative Rods

Facultatively anaerobic gram-negative rods are a group of bacteria that take on a rod shape and can live without the presence of oxygen or when oxygen is present can carry out metabolism aerobically. These bacteria are gram-negative. These are three prominent members of the facultatively anaerobic gram-negative rods bacteria group.

- *Enterics.* Enterics (*Enterobacteriaceae)* are small bacteria that are found in the intestinal tracts of animals and humans (*intestinal flora*) and have flagella all over their surface (*peritirichous flagella*) to move about. Enterics ferment glucose and produce carbon dioxide and other gases. The word "enteric" means "pertaining to the intestines." More predominate Enterics are:

  - *Escherichia coli. Escherichia coli,* commonly known as *E.coli* is an example of an enteric bacteria, which makes up some of the normal flora in the human intestines, but can cause infection if it enters other parts of the body (for example, through the ingestion of water that is contaminated with fecal matter.)

- *Shigella. Shigella* causes bacillary dysentery or shigellosis, more commonly known as traveler's diarrhea.

- *Salmonella. Salmonella* is a group (genera) of enteric bacteria that has members that can infect humans. Species include *Salmonella typhi*, which causes typhoid fever and *Salmonella choleraesuis* and *Salmonella enteritides* both of which are food-borne pathogens that cause salmonellosis, a type of food poisoning.

- *Klebsiella.* The genera *Klebsiella* is a bacterium that cause bacterial pneumonia.

- *Erwinia. Erwinia* is a bacterium that infects plants and causes what is commonly called soft root rot.

- *Enterobacter.* The genus *enterobacter* consists of the species *Enterobacter cloace* and *Enterobacter aerogenes*. Both of these organisms cause urinary tract infections and nosocomial (hospital) infections in individuals with a weakened immune system.

- *Serratia marcescens, Serratia marcescens* bacteria arc found on catheters and instruments that are allegedly sterile; this bacterium causes urinary and respiratory infections.

- *Yersinia pestis. Yersinia pestis*, also known as *Y.pestis*, caused the bubonic or black plague, which ravaged Europe during the Middle Ages. It begins by causing abscesses of lymph nodes and then produces pneumonia-like symptoms when it reaches the lungs, which is called pneumonia plague.

- Vibrios. Vibrios are facultative anaerobic gram-negative, comma-shaped bacteria that inhibit aquatic environments and some also live in the intestinal tracts of animals and humans. There are two important species of vibrios bacteria. These are:

  - *Vibrio cholerae. Vibrio cholerae* is the bacteria that cause cholera, signs and symptoms are abdominal pain and watery diarrhea.

  - *Vibrio parahaemolyticus. Vibrio parahaemolyticus* is the bacterium that causes inflammation and irritation of the stomach and intestine— better known as gastroenteritis—when contaminated shellfish is ingested raw or undercooked.

- *Pasteurella-Haemophilus. Pasteurella-Haemophilus* are very small facultatively anaerobic gram-negative, rod shaped bacteria, that are named for Louis Pasteur. Examples include:
  - *Pasteurella* causes blood poisoning (septicemia) in cattle, chickens (fowl cholera) and pneumonia in various animals.
  - *P. multocida,* the species which was identified by Louis Pasteur, is transmitted to humans by a dog or cat bite.
  - *Haemophilus.* The genus *Haemophilus* (which means "blood loving") are bacteria that live in mucous membranes of the upper respiratory tract, mouth, intestinal tract and vagina.
  - *H. ducreyi*, the species of *Haemophilus* bacterium that causes *chancroid*, which is an infectious venereal ulcer.
  - *H. aegyptiusis* the species that causes acute conjunctivitis or pink eye.

Anaerobic Gram-Negative Cocci and Rods

Anaerobic gram-negative cocci and rods can live in anaerobic conditions and are non-motile. They do not form endospores, which are small spores that developed inside the bacteria as a resistant survival form of the bacteria.

Anaerobic gram-negative cocci are spherically shaped and form a single chain or are clustered. *Veillonella* are common anaerobic gram-negative cocci that are found between teeth and on gums. *Veillonella* is the cause of abscesses of teeth and gums.

Anaerobic gram-negative rods are called *Bacteroides* and are members of the *Bacteriodaceae* family of bacteria that live in the intestinal tract of humans. Bacteroides can cause peritonitis, which is inflammation of the peritoneum due to infection. Another kind of anaerobic gram-negative rod is *Fusobacterium*. These are long slender rods that live in the gingival crevices of teeth and cause gingivitis, which is a gum infection.

Rickettsias and Chlamydias

Rickettsias and Chlamydias are intracellular parasites that need a host in order to reproduce and therefore enter the cell of a host. These bacteria were once thought to be viruses that invaded cells. They are classified as bacteria because they have bacterial cell walls and contain DNA and RNA, which is not the case with a virus. Rickettsias and Chlamydias have no means of mobility because they lack flagella. They are also gram-negative.

Rickettsias are small rod-shaped or spherical bacteria that live in the cells of ticks, lice, fleas, and mites (arthropods) and can be transmitted to humans when bitten by arthropods, causing rickettsial disease. *Rickettsial disease* causes capillaries to become permeable resulting in a rash. Rickettsias reproduce by binary fission where a cell wall forms across the cell and the two halves separate to become individual cells.

Common Rickettsias are:

- *Rickettsia prowazekii. Rickettsia prowazekii* is transmitted by lice and causes endemic typhus.
- *R. rickettsii. R. rickettsii* is transmitted by ticks and causes *Rocky Mountain spotted fever.*
- *R. tsutsugamushi. R. tsutsugamushi* is transmitted by arthropods and cause *scrub typhus* that presents with fewer, rash and inflammation of the lymph nodes.
- *Coxiella burnetti. Coxiella burnetti* is transmitted by aerosols or contaminated milk and causes Q fever, which is similar to pneumonia.
- *Bartonella bacilliformis. Bartonella bacilliformis* is transmitted by arthropods and causes a wart-like rash called *Oroya fever.*

Chlamydias are very small spherical or coccoid bacteria that are non-motile and can be transmitted from person-to-person contact or by airborne respiration. Chlamydias are not transmitted by arthropods. They are so small that they multiply in host cells. There are three species of chlamydia:

- *Chlamydia trachomatis. C.trachomatis* causes trachoma, which is a common cause of blindness and the non-gonorrhea sexually transmitted disease urethritis (inflammation of the urethra).
- *C. penemoniae* causes a mild form of pneumonia in adolescence.
- *C. psittaci* causes *psittacosis.*

Mycoplasmas

Mycoplasmas are very small facultatively anaerobic bacteria (some are obligately anaerobic) that have taken on many shapes (*pleomorphic*) and were once

thought to be viruses because they lack a cell wall. However, they have a cell membrane, DNA and RNA, which distinguishes them from viruses.

Mycoplasmas can also resemble fungi because some Mycoplasmas produce filaments that are commonly seen in fungi. It is these filaments that led scientists to name it Mycoplasma. Myco means "fungus."

Many Mycoplasmas are unable to move by themselves because they do not have flagella, but some are able to glide on a wet surface.

Two of the more common types of Mycoplasma are:

- *Mycoplasma pneumoniae. Mycoplasma pneumoniae* is the cause of atypical pneumonia, commonly referred to as walking pneumonia.

- *Ureaplasma urealyticum. Ureaplasma urealyticum* is a bacterium that is found in urine and one that can cause urinary tract infection.

Gram-Positive Cocci

Within the gram-positive cocci section are two genera. These are *Staphylococcus* and *Streptococcus,* and each has an important role in medicine.

- *Staphylococcus. Staphylococcus* bacteria have a grape-like cluster appearance and grow in environments of high osmotic pressure and low moisture. Osmotic pressure is the pressure required to prevent the net flow of water by osmosis. Infections caused by the Staphylococcus bacteria are typically called *staph* infections. Here are the commonly found Staphylococcus bacteria:

  - *Staphylococcus aureus. Staphylococcus aureus*, also called S. aureus, is a bacterium that forms yellow-pigmented colonies that grows with oxygen (aerobically) or without oxygen (anaerobically). *S. aureus* is the cause of toxic shock syndrome that results in high fever, vomiting and sometimes death. It produces *enterotoxins*, which affects intestinal mucosa. *S. aureus* is also the cause of boils (skin abscess), impetigo (pus-filled blisters on the skin), styes (an infection at the base of an eye lash), pneumonia, osteomyelitis, acute bilateral endocarditis (inflammation of the internal membranes of the heart) and scalded skin syndrome in very young children that causes skin to strip off (denude) due to an exfoliative toxin.

    S. aureus is one of the major types of infections that occur in hospitals because it is resistant to antibiotics such as penicillin. An infection of *S. aureus* is usually identified by the presence of an abscess.

- *Staphylococcus epidermis*: *Staphylococcus epidermis*, also known as *S. epidermis*, is the frequent cause of urinary tract infections in the elderly and also causes subacute bacterial endocarditis, which is a chronic infection of the endocardium (a thin layer of connective tissue that lines the chambers of the heart) and heart valves.
- *Staphylococcus saprophyticus*. *Staphylococcus saprophyticus*, also known as *S. saprophyticus*, causes urinary tract infections, usually in adolescent girls.

- *Streptococcus*. Streptococcus bacteria appear as a single, paired or chained spherical gram-positive bacteria. Streptococci do not use oxygen, though most are aerotolerant. Few may be obligately anaerobic. Infections caused by the Streptococcus bacteria are generally referred to as a *strep* infection. Microbiologists classify Streptococcus bacteria in three ways.
  - The type of hemolysis (destruction) of red blood cells caused by the Streptococcus bacteria. There are three types characterized by:
    - Alpha-hemolytic group. Incomplete *lysis* (destruction of the cell) within green pigment surrounding the colony.
    - Beta-hemolytic group. Total lysis and a clear area around the colony.
    - Gamma-hemolytic group. Absence of lysis. This group is of no clincal importance.
  - The Lance Field classification. There are four groups:
    - Group A Streptococci. Characterized by *Streptococcus pyogenes* and secrete of erthrogenic exotoxins responsible for scarlet fever.
    - Group B Streptococci. Characterized by *Streptococcus agalactiae*, which is part of normal oral and vaginal flora and causes *urogenital* (urinary and reproductive systems) infections in females.
    - Group C Streptococci. Causes animal diseases.
    - Group D Streptococci. Characterized by *Streptococcus faecalis*, which is a normal part of oral and intestinal flora. Diseases of *S. faecalis* are endocarditis, urinary tract infections and septicemia (blood poisoning).
  - Ungrouped Streptococci. There are two kinds:
    - *Viridans streptococci*. Characterized by *Streptococcus viridans* and *Streptococcus salvarius,* which causes subacute bacterial endocarditis, and *Streptococcus mutans* which causes a biofilm called plague resulting in tooth decay.
    - *Pneumococcal streptococci*. Characterized by *Streptococcus pneumoniae* which causes lobar pneumonia and otitis media (middle ear infection).

Endospore-Forming Gram-Positive Rods and Cocci

Endospore forming, gram-positive rods and cocci consist mainly of rod-shaped bacteria of the genera *Bacillus* and *Clostridium*. Another cocci bacteria included in this group are of the genus *Sporosarcina*. These bacteria have no clinical significance and are saprophytic soil bacteria. Saphrophytes are organisms that feed on dead organic matter.

These bacteria can be strict aerobes, facultative anaerobes, obligate anaerobes or microaerophiles. Microaerophiles are bacteria that grow best in an environment that has a small amount of free oxygen.

The formation of endospores by bacteria is important in medicine and the food industry because these endospores are resistant to heat and many chemicals.

There are three genera in this section. These are:

- Bacillus: Bacillus consists of the following bacterium:

  - *Bacillus anthracis. Bacillus anthracis* causes anthrax, a severe blood infection that infects cattle, sheep and horses and can be transmitted to humans. *B. anthracis* is a non-motile facultative anaerobe and produces exotoxin. Anthrax can result in central nervous system distress, respiratory failure, anoxia and death.

  - *Bacillus cereus. Bacillus cereus* produces enterotoxin (a toxin that affects the intestine) and causes gastroenteritis (food poisoning).

  - *Bacillus thuringiensis. Bacillus thuringiensis* produces a toxin that attacks the digestive system of insects, causing the insects to stop feeding by causing paralysis of the insects guts.

- *Sporosarcina. Sporosarcina* are bacteria that inhabit the soil and receive nutrients by feeding on dead organic matter.

- *Clostridium.* Clostridium are rod-shaped bacteria that exist in water, soil, and in the intestinal tract of animals and humans. These bacteria do not require oxygen. They release toxins that cause disease. Here are the common types of Clostridium:

  - *Clostridium tetani. Clostridium tetani*, also known as *C. tetani*, causes tetanus, commonly referred to as lockjaw.

  - *Clostridium difficile. Clostridium difficile*, also known as *C. difficile*, causes gastroenteritis.

- *Clostridium perfringes*. *Clostridium perfringes*, also known as *C. perfringes*, causes myonecrosis—better known as gas gangrene—which produces hydrogen gas in deep tissue wounds, resulting in cell death.

- *Clostridium botulinum*. *Clostridium botulinum*, also known as *C. botulinum* is a cause of food poisoning usually as a result of improperly canned food. It produces an exotoxin that causes flaccid paralysis (weakness of muscle tone) due to the suppression of acetylcholine, which is a neurotransmitter. The result is vomiting, difficulty speaking, and difficulty swallowing, which can lead to respiratory paralysis and death. Physicians use *Clostridium botulinum* as a neural block that inhibits muscle contraction. *Clostridium botilinum* is also used cosmetically to relax muscles that cause facial wrinkles (Botox injections). The *C. botulinum* toxin blocks the exocytosis of synaptic vesicle of the neuromuscular junction, where motor neurons meet muscle.

Regular Nonsporing Gram-Positive Rods

Regular, non-sporing gram-positive rods are obligate or facultative anaerobes that have a rod-shaped appearance as is implied by its name. They inhabit fermenting plants and animal products. There are four genera within this section. These are:

- The genus, *Lactobacillus* are non-sporing gram-positive rods. These are aerotolerant bacteria that produce lactic acid from simple carbohydrates. The acidity inhibits competing bacteria. An example of a *Lactobacillus* organism is the species *Lactobacillus acidophilus*.

- *Lactobacillus acidophilus* is found in the human intestinal tract, oral cavity and adult vagina. They produce an acidic environment that inhibits the growth of harmful bacteria by fermenting glycogen into lactic acid. *Lactobacillus acidophilus* is also used commercially to produce an assortment of food products including sauerkraut, pickles, buttermilk and yogurt.

Other examples of regular non-sporing gram-positive rod bacteria are:

- *Listeria monocytogenes*. *Listeria monocytogenes* contaminates food and dairy products, if ingested can cause the disease listeriosis, which causes an inflammation of the brain and meninges (meningitis).

- *Erysipelothrix rhusiopathiae. Erysipelothrix rhusiopathiae* causes erysipeloid, which is red, swollen and painful lesions, frequently seen in fishermen and meat handlers.

Irregular Nonsporing Gram-Positive Rods

These bacteria are irregular, non-sporing rods. Although this group of bacteria are generally rod-shaped, their shape can vary (pleomorphic). Some resemble a club while others are long, threadlike cylinders. There are three genera within this section. These are:

- *Corynebacteria*. Corynebacteria are club-shaped and receive nutrients from dead or decaying organic material (*saprophytes*). Corynebacteria inhabit airy soil and water and cause diphtheria. *Corynebacteria diphtheriae* is the organism which causes diphtheria.
- *Propionibacterium*. Propionibacterium infects wounds and causes abscesses. An example would be *Propionibacterium acnes*.
- *Actinomycetales*. Actinomycetales is a long, threadlike cylinder (filament) that inhibits soil and some provide nitrogen to plants. The species *Actinomycetale israelii*, which causes *actinomycosis*, which destroys tissues in the jaw, head, neck, and lungs. Actinomycetales was originally classified as a fungus because of its shape.

Mycobacteria

Mycobacteria require oxygen (aerobic) and are acid-fast organisms that remain red while most are blue. Large amounts of lipids in the Mycobacteria's cell envelope, resists basic dyes. Myco, which means "fungus-like" is how this organism got its name.

- *Mycobacterium tuberculosis. Mycobacterium tuberculosis* causes tuberculosis.
- *Mycobacterium leprae. Mycobacterium leprae*, also known as *M. leprae*, causes Hansen's disease (leprosy).
- *Mycobacterium bovis. Mycobacterium bovis*, also known as *M. bovis*, causes tuberculosis in cattle and can be transmitted to humans.

Nocardia Forms

Nocardia are a group of a long thread-like cylinder shaped bacteria that inhabits soil and need oxygen to grow (aerobic). They are gram-positive and cannot move by themselves (they are non-motile).

Nocardia asteroids, also known as *N. asteroids*, is a common bacterium within this group. It causes *mycetoma*, which causes abscesses on the hands and feet and also causes lung infection.

Quiz

1. Archaea can exist in very hot and very cold environments, making them a resilient microorganism that can survive attacks and destroy other bacteria.
   (a) True
   (b) False

2. Archaea that function in the presence of air are called
   (a) methanogenic archaea
   (b) cell wall–less archaea
   (c) sulfate reducers
   (d) extreme halophiles

3. Archaea that live in an extreme salty environment are called
   (a) methanogenic archaea
   (b) cell wall–less archaea
   (c) sulfate reducers
   (d) extreme halophiles

4. Single-celled archaea that produce methane and carbon dioxide (CO_2) by fermenting simple organic carbon compounds or oxidizing H_2 without oxygen to produce CO_2 are called
   (a) methanogenic archaea
   (b) cell wall–less archaea
   (c) sulfate reducers
   (d) extreme halophiles

5. Archaea that do not have cell walls are called
   (a) methanogenic archaea
   (b) cell wall–less Archaea
   (c) sulfate reducers
   (d) extreme halophiles

6. "Aerobic/microaerophilic" is a designation for bacteria that require small amounts of oxygen to grow.
   (a) True
   (b) False

7. Rod-shaped bacteria with flagella at both ends of the rod, which give the bacteria mobility, are called
   (a) *Pseudomonas aeruginosa*
   (b) *Legionella pneumophilia*
   (c) pseudomonads
   (d) *Legionella micdadei*

8. The bacteria that infects lungs and causes a strain of pneumonia commonly called Pittsburgh pneumonia is known as
   (a) *Pseudomonas aeruginosa*
   (b) *Legionella pneumophilia*
   (c) pseudomonads
   (d) *Legionella micdadei*

9. A pathogenic bacterium that infects the urinary tract and wounds in humans is called
   (a) *Pseudomonas aeruginosa*
   (b) *Legionella pneumophilia*
   (c) pseudomonads
   (d) *Legionella micdadei*

10. This bacterium, discovered in 1976, infected and killed members of the American Legion at their convention in Philadelphia.
    (a) *Pseudomonas aeruginosa*
    (b) *Legionella pneumophilia*
    (c) pseudomonads
    (d) *Legionella micdadei*

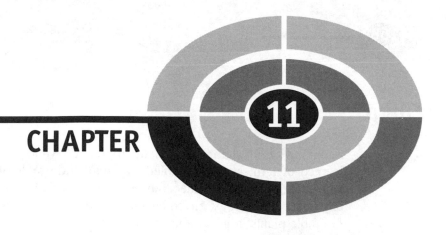

CHAPTER 11

The Eukaryotes: Fungi, Algae, Protozoa, and Helminths

Throughout this book you've learned that some kinds of microorganisms are beneficial to us: They supply food, remove waste, and help prevent disease by combating bacteria as antibiotics. Other microorganisms however, are pathogenic; they cause or transmit disease.

In this chapter we take a close look at microorganisms called eukaryotes. Eukaryotes are organisms within the kingdoms Fungi, Plants, Protists, and Animals. These microorganisms are called fungi, algae, protozoa, and helminths.

Fungi

Fungi have been studied systematically for 250 years, although ancient peoples learned of fermentation (enabled by fungi) thousands of years ago. Scientists who practice *mycology*, the study of fungi, are called *mycologists*. In the early days of microbiology, mycologists categorized fungi as plants because they resemble plants in general appearance (they have cell walls) and because both fungi and plants lack motility (neither can move under its own power).

Today, however, fungi and plants are considered two distinct groups of organisms because plants use chlorophyll to obtain nutrients and fungi do not. Fungi are heterotrophic: They absorb nutrients from organic matter and organic wastes (*saprophytes*) or tissues of other organisms (*parasites*). Many fungi are multicellular and are called molds. Yeasts are unicellular fungi.

Fungi can be both beneficial and harmful. For example, fungi called *mycorrhizae* are mutualistic and help roots of plants absorb water and minerals from the soil. The cellulose and lignin of plants are important food sources for ants; however, ants are unable to digest them unless fungi first break them down. Ants are known to cultivate fungi for that purpose. Some fungi are beneficial to humans as food (mushrooms). They are used in the preparation of food such as bread and beer (yeast). Fungi are also used to fight off bacterial diseases (antibiotics).

Some fungi can have a harmful effect because they feed on plants, animals, and humans, causing plants and animals to decay and spoil as a source of nutrients (rotting food). In humans, fungi cause various diseases such as athlete's foot.

ANATOMY OF FUNGI

The body of a fungus (Fig. 11-1) is referred to as either a *soma* (meaning "body"), which is equal to the term "vegetative" in plants, or *thallus,* which is also applied to algae and bryophytes (nonflowering plants comprised of mosses and liverworts). The body of a mold or fleshy fungus consists of long, loosely packed filaments called *hyphae.*

Hyphae are divided by cell walls called *septa* (the singular form is *septum*). In most molds the hyphae are divided into one cell units called *septate hyphae.* In some fungi, the hyphae have no septa and look like long multinucleated cells called *coenocytic hyphae.* Cytoplasm flows or streams throughout the hyphae through pores in the septa. Under the right environmental conditions the hyphae grow to form a *filamentous mass* known as a *mycelium.* A fungus can

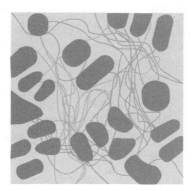

Fig. 11-1.   The body of a fungus contains
long filaments called hyphae.

have a thallus many meters and penetrate its surroundings. In the hyphae of fungi
there is a portion called the vegetative hyphae. Vegetative hyphae are where nutrients are obtained. The part of the hyphae responsible for reproduction is called
reproductive or aerial hyphae.

Fungi can reproduce both sexually and asexually. Reproduction occurs with
the formation of spores. Spores are always *nonmotile* and are a common means
of reproduction among fungi. Do not confuse bacterial endospores with fungal
spores; they are different. Bacterial endospores are formed so that the bacterial
cell can survive in harsh environments. Once there is a less threatening en-
ironment, the bacterium leaves the endospore state and becomes active. The
endospore germinates into a single bacterial cell. *Asexual reproduction* occurs
when *asexual spores* are formed by the hyphae of one organism. When these
spores germinate, they are identical to the parent. *Sexual reproduction* happens
when the nuclei of *sexual spores* from two opposite mating strains of the same
fungus species fuse. Fungi that grow from sexual spores have genetic characteristics of both parents.

YEASTS

Yeasts are fungi that are unicellular and reproduce using a process called fission,
although some can form filaments. *Fission* occurs when a cell divides evenly to
form two new cells. When the cell divides by budding, it divides unevenly.
Yeasts are nonfilamentous and have a spherical or oval shape. The white pow-

dery substance that is sometimes found coating fruits and leaves is a yeast. Yeasts can also reproduce sexually.

MOLDS

When a mold forms an asexual spore, the spore will detach itself from the parent and then germinate into a new mold. This process is considered reproduction because a second new organism grows from the spore.

FUNGI CLASSIFICATION

Asexual fungal spores are formed on hyphae of fungi. When these hyphal spores germinate, they are identical to the parent. Asexual spores reproduce by the process of cell division. In sexual cell reproduction, the spores are produced by the fusion of nuclei from two opposite fungi of the same species. These fungi will have the same characteristics of both parents. Asexual spores produce more frequently than sexual spores. Asexual spores are present in virtually every environment on the planet.

Some fungi change their structure based on their natural habitat. This is referred to as dimorphism, the property of having two forms of growth. For example, some fungi appear non-filamentous when growing outside their natural habitat but filamentous when growing in their natural habitat. Such changes of appearance can make it challenging to identify a particular type of fungus. Fungal classification is based on the type of sexual spores they produce.

Listed are examples of the divisions of the kingdom Fungi:

- *Zygomycota: Zygomycota* are conjugative fungi. They reproduce both sexually (zygospores) and asexually (sporangiospores). An example is: *Rhizopus nigricans*, a black bread mold.

- *Ascomycota: Ascomycota*, also called *sac fungi*, have sac-like cells called *asci*. These are yeasts, truffles, morels, and common molds. Fungi in this group reproduce sexually and asexually. Their sexual spores (*conidiosphores*) freely detach with the slightest movement (*conidia*) and therefore can cause infection (opportunistic disease) or an allergic reaction. Examples are:

  - *Blastomyces*: *Blastomyces* causes *blastomycosis*, which is a general pulmonary disease.

  - *Histoplasma*: *Histoplasma* is a fungus found in bird and bat droppings; it causes *histoplasmosis*, which is known as the *fungus flu*.

- *Basidiomycota:* Also called *club fungus*, basidiomycota includes mushrooms, toadstools, smuts, and rusts. Sexually produced *basidiospores* are formed externally on a base pedestal, producing a club-shaped structure called a *basidium* (*basidia*, plural). Basidia can be found on visible fruiting bodies called *basidiocarps*, which are positioned on stalks. A mushroom is a basidiocarp. Some mushrooms, such as *Amanita,* produce toxins and are poisonous to humans, while others are very nutritious.

- *Deuteromycota: Deutermycota*, also known as *fungi imperfecti*, have no sexual reproduction (or none that can be observed). They cause *pneumocystis* which infects people who have a compromised immune system. Examples are:
  - *Penicillium notatum*, which produces penicillin.
  - *Candida albicans*, which causes vaginal yeast infections in humas.

FUNGUS NUTRITION

Fungi receive nutrients by absorptive nutrition (*chemoheterotrophic*), which is somewhat similar to how bacteria obtain nutrients. Fungi team up with bacteria to break down organic molecules and are the principal decomposers on earth. Fungi can metabolize complex carbohydrates, such as the lignin in wood.

Fungi can decompose substances that have very little moisture and substances that live in an environment with a pH of 5. Almost all molds are aerobic and most yeasts are facultative anaerobes.

Algae

Algae are very simple unicellular or multicellular eukaryotic organisms that obtain energy from sunlight (*photoautotrophs*). They live in various water environments (oceans and ponds) on moist rocks and trees, and in soil.

REPRODUCTION OF ALGAE

Sexual reproduction occurs in most species of algae. In these species the algae reproduce asexually for generations until there is a change in environmental conditions; then the algae reproduce sexually. Other types of algae alternate in how they reproduce. The algae that reproduce sexually will latter reproduce asexually. All algae reproduce asexually. Unicellular algae divide by mitosis and cytokinesis. Multicellular algae that contain thalli and filaments can fragment. Each new piece can form a thallus and a filament.

TYPES OF ALGAE

Chrysophytes

Chrysophytes are unicellular algae that live in fresh water and contain *chlorophyll a* and *chlorophyll c*, which are *photosynthetic* pigments used to transform sunlight into energy. These are also known as golden algae because they have golden silica scales. There are 500 known species of chrysophytes. Some chrysophytes are amoeboid that attack bacteria by engulfing and destroying it.

Diatoms

Diatoms are unicellular algae that have a hard, double outer shell made of silica. Nutrients pass through pores in the shell, then through the diatom's plasma membrane contained within the shell. There are 5,600 known species of diatoms, most of which are phototrophic and contain chlorophyll a and chlorophyll c pigments. They also contain *carotenoids*, which are yellow and orange pigments. Some diatoms are *heterotrophs* and break down and use organic matter as nutrients. Diatoms accumulate at the bottom of the sea and are commercially mined for both their value as an abrasive and their filtering and insulating capabilities (used in the filters in pools).

Dinoflagellates

Dinoflagellates are unicellular algae that have the capability of self-movement through the use of tail-like projections called *flagella*. The flagella are located between grooves in the cellulose plates that cover the dinoflagellate's body. These flagella pulsate in both an encircling motion around the body and in a perpendicular motion, causing the dinoflagellates to rotate like a top. There are about 1,200 known species of dinoflagellates that inhabit both fresh water and seawater.

Dinoflagellates live in seawater. Some are heterotrophs and break down organic matter for nutrients. Some seawater dinoflagellates are luminous, giving a twinkle to the sea at night. Freshwater dinoflagellates are phototrophic: They synthesize nutrients from sunlight using photosynthesis.

Many dinoflagellates have chlorophyll a and c pigments as well as the yellow and orange pigments, carotenoids. Depending on their photosynthetic pigment, dinoflagellates can appear yellow-green, green, brown, blue, or red. When dinoflagellates undergo a population explosion, the sea changes color from an ocean blue to a sea of red or brown.

Some dinoflagellates, such as *Gonyaulax (plankton),* can be fatal to humans because they produce neurotoxins. These dinoflagellates are eaten by fish and are absorbed by oysters, clams, and other shellfish (mollusks). The neurotoxins build up in their tissues, making the seafood poisonous to humans. These dinoflagellates also cause "red tides" that have a devastating effect on the fish population.

Red Algae

Red algae, also known as *Rhodophyta,* are algae that form colonies in warm ocean currents and in tropical seas. They contribute to the formation of coral reefs that can be found as deep as 268 meters below the surface of the ocean. Their stone-like appearance is caused by a build-up of calcium carbonate deposits on their cell walls.

There are 4,000 known species of red algae, of which fewer than 100 are found in fresh water. Red algae get their color from the phycobilins and chlorophyll a pigments contained in their cells. *Phycobilins* pigment absorbs green, violet, and blue light, which are light waves that are capable of penetrating the deepest waters. It is for this reason that red algae can survive at great depths. The pigment that makes the algae red is called phycoerythrin.

As you learned in Chapter 6, red algae are used to make agar. Agar is the culture medium that is extracted from the cell wall of red algae and is used to grow bacteria. Red algae are also used as a moisture-preserving agent in cosmetics and baked goods. Red algae are used as a setting agent for jellies and desserts.

Brown Algae

Brown algae, also known as *phaeophyta*, are multicellular organisms. Some brown algae are commonly called *kelp;* they live in the northern rocky shores of North America and can grow up to 30 meters. There are 1,500 known species of brown algae.

Brown algae have chlorophyll a and b photosynthetic pigments. They also have carotenoids. Brown algae can appear dark brown, olive-green, and even golden depending on the type of pigments in their cells. The pigment that makes the algae brown is called *fucoxanthin. Algin* is a gummy substance found in the cell walls of some species of brown algae and is used as a thickening, emulsify-

ing, and suspension agent for ice cream, pudding, frozen foods, toothpaste, floor polish, cough syrup, and even jelly beans.

The organic matter that kelp produces supports the life of invertebrates, marine mammals, and fish.

Green Algae

Green algae can live in moist places on land, such as tree trunks and in the soil, as well as in water. There are 7,000 species of green algae that are diverse in size, morphology, lifestyle, and habits. Scientists believe that some members of the species are linked structurally and biochemically to the Plant kingdom. Two common green algae are:

1. *Spirogyra: Spirogyra* are freshwater algae that have tiny filaments, each containing spiraling bands of chloroplasts.
2. *Volvox: Volvox* are colonial multicellular green algae that have flagella and live in marine, brackish, and freshwater environments.

LICHENS

Lichens are filaments of a fungus and cells of algae (this is a symbiotic relationship) that are found on exposed soil or rock, on trees, on rooftops, and on cement structures. There are about 20,000 known species of lichens.

Survival of the green algae and the fungus are interdependent in a symbiotic association. Neither can live without the other. However, each grows independently. Lichens are delicate and beautiful in appearance.

Protozoa

This organisms are members of the Kingdom Protista. There are about 20,000 known species of protozoa that live in water and soil. Some feed on bacteria while others are parasites and feed off their hosts.

Most protozoa are asexual and reproduce in one of three ways. These are:

- *Fission:* Fission occurs when a cell divides evenly to form two new cells.
- *Budding:* Budding occurs when a cell divides unevenly.

- *Multiple fission (schizogony):* Multiple fission is when the nucleus of the cell divides multiple times before the rest of the cell divides. Forms around each nucleus when the nucleus divides then each nuclei separates into a daughter cell.

Some protists are sexual and exchange genetic material from one cell to another through *conjugation*, which is the physical contact between cells.

A protist can survive in an adverse environment by encapsulating itself with a protective coating called a *cyst*. The cyst defends the protist in extreme temperatures against toxic chemicals and even when there is a lack of oxygen, moisture, and food.

PROTOZOA NUTRITION

Protists receive nutrients by breaking down organic matter (*heterotrophic*) and can grow in both aerobic and anaerobic environments, such as protists that live in the intestine of animals. Some protists, such as *Euglena*, receive nutrients from organic matter and through photosynthesis because they contain chlorophyll. These protists are considered both algae and protozoa.

Protists obtain food in one of three ways:

- Absorption: Food is absorbed across the protist's plasma membrane.
- Ingestion: Cilia outside the protist create a wave-like motion to move food into a mouth-like opening in the protist called a *cytosome*. An example is the paramecium.
- Engulf: *Pseudopods* (meaning "false feet") on the protist engulf food, then pull it into the cell using a process called *phagocytosis*. An example of this type of protist is the amoeba.

Food is digested in the vacuole after the food enters the cell. The *vacuole* is a membrane-bound organelle. Waste products are excreted using a process called *exocytosis*.

AMEBA

Here are common amebas (Fig. 11-2):

- *Entamoeba histolytica:* Also known as *E. histolytica,* this microorganism is transmitted between humans through the ingestion of cysts that are

Fig. 11-2.   A common type of ameba is the *Amoeba proteus*.

excreted in the feces of infected people. It is the causative agent of amebic dysentery.

- *Naegleria fowleri:* This ameba causes primary amebic meningoencephalitis (PAM) that results in headache, fever, vomiting, stiff neck, and loss of bodily control. *N. fowleri* enters the body through the mucous membranes (when the person swims in warm water) and travels to the brain and spinal cord.

- *Acanthamoeba polyphaga:* This ameba lives in water (including tap water) and infects the cornea of the eye leading to blindness. It can also cause ulcerations of the eye and the skin. *A. polyphaga* is also known to invade the central nervous system, resulting in death.

Flagellates move by structures called flagella. They have two or more spindle-shaped flagella in the front of the cell that they use to pull themselves through their environment. Food enters flagellates through a mouth-like grove called a *cytosome*.

Here are common flagellates:

- *Trichomonas vaginalis: C*ommonly known as *T. vaginalis*, this flagellate is the cause of *trichomoniasis*, which is a sexually transmitted disease. *T. vaginalis* is found in the male urinary tract and the vagina of females.

- *Giardia lamblia:* This flagellate is commonly known as *G. lamblia* and causes giardiasis. *Giardiasis* causes nausea, cramping, and diarrhea when food or water contaminated by fecal material is ingested. *G. lamblia* lives in the small intestines of humans and other mammals.

BLOOD AND TISSUE PROTOZOA

Hemoflagellates are protozoa that are carried by blood-feeding insects and are transmitted into the blood stream by the insect's bite. Here are commonly found hemoflagellates:

- *Trypanosoma gambiense: T. gambiense* is transmitted in the saliva of the tsetse fly as a result of a bite and causes *trypanosomiasis*, which is better known as *African sleeping sickness.*

- *Trypanosoma cruzi: T. cruzi* is carried by the Reduviid bug or "kissing bug," which is commonly called the kissing bug because it bites the face. *T. cruzi* causes *Chagas' disease.* Chagas' disease is thought to have made Charles Darwin sick during his voyage on the *H.M.S. Beagle.* Typically there aren't any symptoms for months after the bite. During this time *T. cruzi* spreads through organs of the body, weakening the heart, intestines, and esophageal. It also cause, both eyes to swell (*Romaña's sign*).

CILIATES

Ciliates (Fig. 11-3) are protozoa that have shorter hair-like structures called cilia that are found in rows on the outer surface of the cell. These cilia are used to move the protozoa through the environment and are used to bring food into the cell.

An example of a ciliate is *Balantidium coli.* It is the only ciliate that causes disease in humans. When ingested, it enters the large intestine, causing severe dysentery.

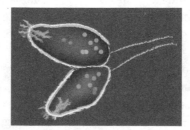

Fig. 11-3.   Ciliates are protozoa that have cilia.

APICOMPLEXANS

Apicomplexans are protozoa that live and grow inside another living organism (obligate intracellular parasites) and cannot move by themselves (nonmotile). Apicomplexans have an apical complex of organelles that form an apex (tip). This tip contains enzymes that enable it to penetrate the tissues of a host. Here are common Apicomplexans:

- *Plasmodium: Plasmodium* lives in the female *Anopheles mosquito* and causes *malaria* when the mosquito bites a human. Symptoms of malaria include severe chills and fever or "rigor" (a sudden chill or coldness that is followed by fever).

- *Bebesia microti: B. microti* lives in ticks and causes *bebesiosis* when the tick bites someone. *B. microti* then enters red blood cells where it multiplies quickly. At first there aren't any symptoms (*asymptomatic*). However, soon afterwards there is a high fever, headache, and muscle pain as *B. microti* destroys red blood cells. This causes the person to become *anemic* (insufficient hemoglobin because of the reduction in the number of red blood cells) and show signs of *jaundice* (an increase in bile, causing the skin and eye sclera—the white part of the eye—to yellow).

- *Toxoplasma gondii*, also known as *T. gondii*, lives in cat feces and raw meat and causes lymphadenitis (infection of the lymph nodes), which can have a devastating effect on people who are immunocompromised, such as AIDS patients. *T. gondii* causes congenital infections in a fetus because it can pass from the mother to the fetus through the placenta.

Helminths

Helminths are parasitic, multicellular eukaryotic animals. The majority of these animals belong to the phyla Platyhelminthes and Nematoda. There are free-living members of these phyla; however, in this section, only the parasitic organisms are discussed.

Many parasitic helminths do not have a digestive system and instead absorb nutrients from the food that is consumed by their host organism, the host's body fluids, and the host's tissues. Parasitic helminths have a very simplistic nervous system because they have to respond to very few changes in their host's environment. They lack or have reduced means of locomotion because they are transferred

from one host to another. Parasitic helminths have a very complex reproductive system that produces fertilized eggs (zygotes) that infect the host organism.

LIFE CYCLE OF HELMINTHS

The life cycle of parasitic helminths that go through a developmental larval stage involves an intermediate host.

Dioecious adult helminths are of one sex. That is, one individual has a male reproductive system and another has a female reproductive system. When these two adult helminths with different sex organs occupy the same host organism, sexual reproduction can occur. *Monoecious* adult helminths are hermaphroditic (an organism that has both female and male reproductive organs). Some of these monoecious helminths can fertilize themselves while others may fertilize each other.

FLATWORMS

Flatworms, also known as platyhelminths, are mostly parasitic, aquatic organisms that range in size from 1 millimeter to 10 meters, as in the case of a tapeworm. There are more than 15,000 known species of flatworms. A flatworm has both male and female reproductive parts (monoecious). Most but not all of their oxygen and nutrients is absorbed through their body wall.

There are two types of flatworms:

- *Flukes:* Flukes are flat, leaf-shaped bodies that have an oral and a ventral sucker that are used to hang on to the body of a host. Flukes live is inside the intestines or on tissues of humans. Three common flukes are:

  - *Schistosoma:* This genus of flukes causes the disease schistosomiasis, a debilitating disease that causes portal hypertension and liver cirrhosis.

  - *Paragonimus westermani: P. westermani* causes *paragonimiasis,* which is the result of the fluke's depositing eggs into the bronchi of the lung.

  - *Clonorchis sinensis: C. sinensis,* also known as *Chinese liver fluke,* causes *clonorchiasis,* which occurs when the fluke latches inside the liver.

- Tapeworms: Tapeworms have a knob-like "head," called a *scolex,* with hooks that allow it to attach to the wall of the intestine of vertebrate animals (including humans). Tapeworms have a series of flat, rectangular body units called *proglottids* (compartments that contain reproductive organs). Proglottids eventually break away from the tapeworm and are excreted in feces. However, new proglottids take their place. A tapeworm continues to grow as long as its scolex and neck are intact.

- *Taenia saginata: T. saginata*, also called beef tapeworm, lives in raw or poorly cooked beef and can cause *taeniasis. T. saginata* can grow to a length of 25 meters in the intestines of a human, leading to an intestinal blockage and malnutrition as the tapeworm absorbs nutrients intended for the person.

- *Tania solium: T. solium*, also called the pork tapeworm, lives in raw or poorly cooked pork and can cause *taeniasis. T. solium* can grow to a length of 7 meters in the intestines leading to an intestinal blockage and malnutrition.

- *Echinococcus granulosus: E. granulosus* is a tapeworm that is spread to humans through contact with an infected dog and is transmitted when a dog licks a person. This can lead to infection, anaphylactic shock, and death if the tapeworm enters the body. *E. granulosus* can lay eggs that produce cysts called *hydatid cysts* in the lungs, liver, and brain.

- *Hymenolepis nana: H. nana* is a tapeworm that lays eggs in cereals and foods that are contaminated with infected parts of insects. When someone ingests the cereal or food he or she also ingests the tapeworm. The tapeworm then attaches to the intestines, leading to diarrhea, abdominal pain, and convulsions.

- *Diphyllobothrium latum: D. latum* is a broad fish tapeworm that lives in raw or poorly cooked fish. The tapeworm attaches to the intestines of the fish where it then lays eggs. While attached, the tapeworm absorbs large quantities of vitamin B_{12} from the intestine eventually causing the person to develop *vitamin deficiency anemia*. This is also, called *pernicious anemia* because there is insufficient vitamin B_{12} to make red blood cells.

ROUNDWORMS (NEMATODES)

Roundworms are also known as nematodes and live in soil, fresh water, and saltwater. Most of the over 80,000 species of roundworms are parasites and live in plants or animals such as insects. They have a primitive body that consists of a cylindrical tube that has tapered ends and is covered with a thick protective layer called a cuticle.

Common roundworms are:

- *Ascaris lumbrieoides:* A. *lumbrieoides* is a roundworm that is transmitted by contaminated human fertilizer, food, or water. It causes *ascariasis*, which is an infection of the small intestine.

- *Strongyloides stercoralis:* S. *stercoralis* is a roundworm whose larvae penetrates human skin and spread into the small intestine where it causes *strongyloidiasis*, which is an infection of the small intestine.

- *Trichinella spiralis:* T. *spiralis* is a roundworm whose larvae cause trichinosis and live in undercooked meats, mainly pork. These juvenile worms that are in the ingested meat mature in the small intestines of the host organism. The mature females burrow through the wall of the small intestines and release their offspring (juveniles) into the blood of the host, where skeletal muscle is soon infected. It is these juveniles that burrow into the skeletal muscle of the host. The larvae travel to the muscle where they form into a sack (*encyst*), causing muscle pain and fever; this results in a large number of eosinophilic leukocytes (*eosinophilia*). An *eosinophilic leukocyte* is a type of white blood cell that increases with allergies and infections.

- *Wuchereria bancrofti:* W. *bancrofti* is a roundworm that lives in mosquitoes and causes *elephantiasis* when the infected mosquito bites a human. The mosquito injects the larvae into the skin where they then migrate to the lymph nodes, causing blockages.

- *Onchocerca volvulus:* O. *volvulus* is a roundworm that lives in the black fly and causes *river blindness* when the black fly bites a human.

- *Dracunculus medinensis:* D. *medinensis* is a roundworm that lives in lobsters, crabs, shrimps, and other crustacea. When the infected crustacea is ingested, this roundworm's larvae migrate from the person's intestines through the abdominal cavity to subcutaneous tissue where they mature. D. *medinensis* releases a toxic substance that creates a skin ulcer, which is the symptom of *dracunculosis* disease.

- Hookworms are roundworms that have tiny hooks that are used to attach it to a host, which is typically the intestine. Here are some common hookworms:

  - *Necator americanus:* N. *americanus*, also known as the New World hookworm, lives in the lower intestine and is the second most common hookworm infection. Its eggs are passed into the feces. Once it comes into contact with a human, it penetrates the skin and spreads into the heart, lungs, and eventually the small intestine where it grows into an adult. This can lead to severe blood loss and anemia.

- *Ancylostoma duodenale:* A. *duodenale*, also known as the Old World hookworm, is similar to N. *americanus*, but is native to Southern Europe, North Africa, Northern Asia, and parts of western South America.
- *Ancylostoma braziliense:* A. *braziliense* is a hookworm that exists in cats and dogs and causes cutaneous larva migrans, which is also known as creeping eruption. Its eggs are passed in feces and the larvae develop in the soil. The larvae can tunnel into the epidermis of humans and can cause an infection to develop.
- *Ancylostoma caninum:* A. *caninum* is a hookworm that exists in dogs and frequently infects puppies. It eats away at the tissues in the small intestine and sucks blood from the dog. This can result in diarrhea, weight loss, anemia, and death. The larvae can tunnel into the epidermis of humans and can cause an infection to develop.

PINWORMS

Pinworms can be up to 10 millimeters long and live in the large intestine. Female pinworms crawl out the anus to lay eggs on the perianal skin. Afterwards, she dies. Pinworms infect about 10 percent of humans, although the person may not know he or she is infected: Pinworms cause few or no symptoms besides a mild gastrointestinal upset and perianal itching, which can lead to bacterial infections. Pinworms are highly contagious and can be transmitted in bed linens and clothing that has been contaminated with eggs. A common pinworm is *Enterobius vermicularis*. Also known as E. *vermicularis*, it is the most common pinworm in the United States. E. *vermicularis* causes *enterobiasis*, in which the skin around the anus is so itchy that a person might not be able to sleep.

QUIZ

1. Helminths are
   (a) fungi
   (b) algae
   (c) protozoa
   (b) worms

2. Fungi are different from plants because
   (a) fungi have chlorophyll
   (b) plants have chlorophyll
   (c) plants absorb nutrients from organic matter
   (d) plants absorb nutrients from organic wastes

3. Mushrooms are
   (a) fungi
   (b) protozoa
   (c) helminths
   (d) algae

4. The body of a mold or fleshy fungus is made up of long, loosely packed filaments called
   (a) soma
   (b) hyphae
   (c) thallus
   (d) mycelium

5. *Fungi imperfecti* can reproduce
   (a) sexually
   (b) do not reproduce
   (c) asexually
   (d) both (a) and (b)

6. These are unicellular algae that have a hard, double outer shell made of silica
   (a) diatoms
   (b) chrysophytes
   (c) dinoflagellates
   (d) phaeophyta

7. What are hemoflagellates?
   (a) Protists
   (b) Nematodes
   (c) Scolex
   (d) Proglottids

8. What are platyhelminths?
   (a) Ringworms
   (b) Nematodes

(c) Flatworms

(d) Hookworms

9. An example of a fluke would be
    (a) *H. ana*
    (b) *D. latum*
    (c) *C. sinensis*
    (d) *E. vermicularis*

10. Some protists use cilia to move food into a mouth-like opening called
    (a) pseudopods
    (b) a cytosome
    (c) a vacuole
    (d) exocytosis

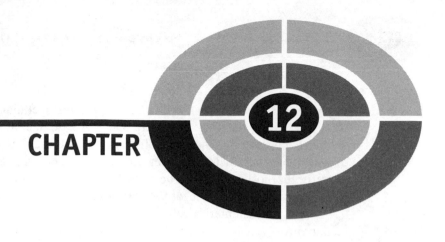

Viruses, Viroids, and Prions

In this chapter, you will learn about viruses, how they work, and the diseases they cause. You will also learn about viroids and prions. A viroid is a small "naked" infectious RNA molecule that has similar properties of a virus. Prions are particles of proteins and cause infections.

Viruses

In 1889, Dutch plant microbiologist Martinus Beijerinck described the concept of viruses through his studies of Tobacco Mosaic Disease. Nobel prize–winner Sir Peter Medawar described how microbiologists feel about viruses when he said, "A virus is a piece of bad news wrapped in a protein."

Viruses are strands of nucleic acids that are encased within a protein coat, making them difficult to destroy. A microorganism needs both DNA and RNA in order to reproduce. A virus cannot express genes without a host, because a virus has either DNA or RNA but not both.

A *virus* (Latin for "poison") is an obligate intracellular parasite that can only replicate inside a living host cell. Once inside the living host cell, a virus becomes integrated in the metabolism of its host, making a virus difficult to control by chemical means. You cannot kill a virus with antibiotics. Drugs that destroy the host's ability to be used by a virus for replication tend to also be highly toxic and have a negatively and sometimes deadly effect on the host cell.

Before a virus enters a cell, it is a free virus particle called a virion. A virion cannot grow or carry out any biosynthetic or biochemical function because it is metabolically inert. Viruses are not cells. They vary in size from 20 nanometers (polio virus) to 300 nanometers (smallpox virus) and cannot be seen under a light microscope.

In 1933, microbiologist Wendell Stanley of the Rockefeller Institute for Medical Research showed that viruses could be regarded as chemical matter rather than as living organisms.

VIRAL STRUCTURE

The major components of a virus are:

- Nucleic acid core. The *nucleic acid core* can either be DNA or RNA that makes up the genetic information (genome) of the virus. RNA genomes only occur in viruses.
- Capsid. A *capsid* is the protein coat that encapsulates a virus and protects the nucleic acid from the environment. It also plays a role in how some viruses attach to a host cell. A capsid consists of one or more proteins that are unique to the virus and determine the shape of the virus.
- Envelope. An *envelope* is a membrane bilayer that some viruses have outside their capsid. If a virus does not have an envelope, the virus is called a naked virus. Examples of diseases caused by naked viruses are chickenpox, shingles, mononucleosis, and herpes simplex. A naked virus is more resistant to changes and is less likely to be affected by conditions that can damage the envelope. Environmental factors that can damage the envelope are:
  - Increased temperature
  - Freezing temperature
  - pH below 6 or above 8
  - Lipid solvents
  - Some chemical disinfectants (chlorine, hydrogen peroxide, and phenol)

Naked viruses are more resistant to changes in temperature and pH. Examples of diseases caused by naked viruses include poliomyelitis, warts, and the common cold.

SHAPES OF VIRUSES

A virus can have one of two structures. These are:

- Helical virus. A *helical virus* is rod- or thread-shaped. The virus that causes rabies is a helical virus.
- Icosahedral virus. An *icosahedral virus* is spherically shaped. Viruses that cause poliomyelitis and herpes simplex are icosahedral viruses.

HOW VIRUSES REPLICATE

The easiest way to understand how viruses replicate is to study the life cycles of viruses called *bacteriophages*. Bacteriophages replicate by either a lytic cycle or a lysogenic cycle. The difference in these two cycles is that the cell dies at the end of the lytic cycle and remains alive in the lysogenic cycle.

The first two scientists to observe bacteriophages were Frederic Twort of England and Felix d'Herelle of France in the early 1900s. The name bacteriophage is credited to d'Herelle and means "eaters of bacteria."

Lytic Cycle

The most studied bacteriophage is the T-even bacteriophage. The virions of T-even bacteriophages are big, complex, and do not contain envelopes. The T-even bacteriophages are composed of a head-and-tail structure and contain genomes of double-stranded DNA. The *lytic cycle* of replication begins with the collision of the bacteriophage and bacteria, called *attachment*. The tail of the bacteriophage attaches to a receptor site on the bacterial cell wall. After attachment, the bacteriophage uses its tail like a hypodermic needle to inject its DNA (nucleic acid) into the bacterium. This is called *penetration*. The bacteriophage uses an enzyme called lysozyme in its tail to break down the bacterial cell wall, enabling it to inject its DNA into the cell. The head or capsid of the bacteriophage remains on the outside of the cell wall. After the DNA is injected into the host's bacterial cell's cytoplasm, *biosynthesis* occurs. Here the T-even bacteriophage uses the host bacterium's nucleotides and some enzymes to make copies of the bacteriophage's DNA. This DNA is transcribed to mRNA, which directs the synthesis of viral enzymes and capsid proteins. Several of these viral enzymes catalyse reactions that make copies of bacteriophage DNA. The bacteriophage DNA will then direct the synthesis of viral components by the host cell.

Next *maturation* occurs. Here the T-even bacteriophage DNA and capsids are put together in order to make *virions*.

The last stage is the *release* of virions from the host bacterium cell. The bacteriophage enzyme lysozyme breaks apart the bacterial cell wall (lysis) and the new virus escapes. The escape of this new bacteriophage virus will then infect neighboring cells and the cycle will continue in these cells.

The Lysogenic Cycle

Some viruses do not cause lysis and ultimate destruction of their host cells which they infect. These viruses are called lysogenic phages or temporate phages. These bacteriophages establish a stable, long-term relationship with their host called *lysogeny*. The bacterial cells infected by these phages are called *lysogenic cells*.

The most studied bacteriophage, which multiplies using the lysogenic cycle, is the bacteriophage Lambda. This bacteriophage infects the bacterium *E. coli*.

When the bacteriophage Lambda penetrates an *E. coli* bacterium, the bacteriophage DNA forms a circle. The circle recombines with the circular DNA of the bacteria. This bacteriophage DNA is called a *prophage*.

Every time the bacteria host cell replicates normally, so does the prophase DNA. On occasion, however, the bacteriophage DNA can break out of the prophage and initiate the lytic cycle.

ANIMAL VIRUSES

Animal viruses infect and replicate animal cells. They differ from bacteriophages in the way they enter a host cell. For example, DNA viruses enter an animal host in this way:

- *Attachment.* Animal viruses attach to the host cell's plasma membrane proteins and glycoproteins (host cell receptors).
- *Penetration.* Animal viruses do not inject nucleic acid into the host eukaryotic cell. Instead, penetration occurs by *endocytosis*, where the virion attaches to the microvillus of the plasma membrane of the host cell. The host cell then enfolds and pulls the virion into the plasma membrane, forming a vesicle within the cell's cytoplasm.
- Transcription in the nucleus by host RNA polymerase.
- Translation by host cell ribosomes.
- DNA replication by host DNA polymerase in the nucleus.
- Assembly of viral particles.
- Release from cell by lysis or exocytosis.

RNA Viruses

- Attachment to host cell receptor.
- Fusion with membrane of host.
- Nucleocapsid enters the cytoplasm.
- Transcription in cytoplasm by viral RNA polymerase.
- Translation by host cell ribosomes.
- Assembly of viral particles.
- Release from cell.

VIRUSES AND INFECTIOUS DISEASE

Viruses are classified by the type of nucleic acid they contain, chemical and physical properties, shape, structure, host range, and how they replicate.

DNA viruses

DNA viruses are viruses that have DNA but no RNA. Common DNA viruses are:

- Hepadnaviruses. *Hepadnaviruses* cause serum hepatitis. The hepatitis B virus (HBV) is a common form of this virus that enters the body via hypodermic needles, blood transfusion, or sexual relations. (Hepatitis A, C, D, E, F, and G are not related and are RNA viruses).
- Herpesviridae. *Herpesviridae* causes the herpes virus. There are about 100 forms of herpes viruses including:
  - Herpes simplex virus type I (HSV-1). *Herpes simplex virus type I* causes encephalitis and enters the body through lesions on the lip, skin, or eyes.
  - Herpes simplex virus type 2 (HSV-2). *Herpes simplex virus type 2* is sexually transmitted, affects the genital and lip area, and can lead to carcinomas.
  - Varicella-Zoster Virus (V2V). *Varicella-Zoster virus* causes chickenpox in the acute form and shingles in the latent form. Shingles appears as vesicles along a nerve resulting in severe pain along the course of the nerve.
  - Cytomegalovirus (CMV). *Cytomegalovirus* causes an infection that usually goes unnoticed unless the person's immune system is compromised,

such as in the bodies of AIDS patients or in infants whose immune systems are not fully developed. This virus can be fatal to some infants.

- Epstein-Barr Virus (EBV). *Epstein-Barr virus* causes infectious mononucleosis.

- Papovaviridae—Papovaviruses such as the human papilloma virus (HPV), which causes warts (papillomas) and tumors *(polyomas)*.

- Poxviridae. *Poxviridae* causes pox (pus-filled lesions) diseases such as smallpox.

- Adenoviridae. Adenoviruses cause acute respiratory disease, which is the common cold virus.

RNA virus

An RNA virus is a virus that contains RNA but not DNA. Common RNA viruses are:

- Flaviviridae. *Flaviviruses,* such as the Dengue virus that causes Break Bone Fever, are carried by mosquitoes. Break Bone Fever results in skin lesions, fever, and muscle and joint pain, and is often fatal. Other Flaviviruses include:

  - St. Louis encephalitis virus. *St. Louis encephalitis virus* causes an infection that is not easily recognized. Wild birds and mosquitoes carry St. Louis encephalitis virus. Monkeys carry a form of this virus called yellow fever virus, which is transmitted to humans by mosquitoes and leads to severe liver damage.

  - Hepatitis C virus (HCV). *Hepatitis C Virus* is called *non-A Hepatitis virus* and *non-B Hepatitis virus* and results in chronic infection. Humans contract this virus from needle pricks and blood transfusions.

- Picornavirus. *Picornavirus* such as the *Poliovirus* causes Poliomyelitis and kills motor neurons resulting in weakness and loss of muscle tone *(flaccid paralysis)*. Others include:

  - *Hepatitis A Virus (HAV)*. Hepatitis A virus (HAV) is also known as infectious hepatitis and is transmitted through a fecal-oral route.

  - *Rhinovirus* is the most frequent cause of the common cold. It causes localized upper respiratory tract infections.

- Retroviridae. Retroviruses are a group of RNA viruses that include the following commonly recognized viruses:

  - Human immunodeficiency virus (HIV). *(Lentivirus) human immunodeficiency virus* is a virus that often results in acquired immunodeficiency syndrome (AIDS) This virus kills T-cells. A T-cell is a white blood cell that fights infection and kills spontaneously arising tumors. HIV causes *Kaposi's sarcoma,* which is a rare form of cancer, and *Pneumocystic carinii,* which is an opportunistic infection and causes pneumonia in AIDS patients.

- Human T-cell leukemia virus 1 and 2. *Human T-cell leukemia virus 1 and 2* is the virus that causes acute *T-cell lymphocytic leukemia* and often contains genes that cause cancer *(oncogenic).*

- Togaviruses. *Togaviruses* is a virus, such as *Estera equine encephalitis,* that is mainly transmitted through blood-sucking insects *(arbovirus),* such as mosquitoes. It causes severe encephalitis. Another is the:

  - Rubella virus. The *rubella virus* causes *German measles,* which can be very dangerous if contracted during the first 10 weeks of pregnancy. The *rubella vaccine* is used to weaken the disease producing ability of the rubella virus.

- Orthomytoviruses. *Orthomytoviruses,* such as *influenza viruses A, B, and C,* cause localized infection of the respiratory tract, which is usually not serious unless the infected person is elderly or the person is infected with secondary bacterial pneumonia. Influenza viruses A and B can cause *Guillain-Barré Syndrome,* which is an inflammation of the nerves that are outside the brain and spinal cord (peripheral nerves); it appears 3 to 5 weeks after a person contracts the flu or after the person receives a flu vaccine. Influenza virus B causes *Reyes syndrome,* which is lethal to the liver and the brain and causes a brain disease (encephalopathy) following a mild flu, chickenpox, or the administration of aspirin.

- Paramyxovirus. *Paramyxovirus,* such as the *parainfluenza virus (Sendai virus)* causes croup in infants. Two other types of paramyxovirus are:

  - Mumps virus. The *mumps virus* causes an enlargement of one or both parotid glands and swelling and pain in the testes and ovaries. There is a vaccine to protect humans from the mumps virus.

  - Measles virus. The *measles virus,* which is also known as *rubeola,* causes measles. The measles virus causes a slow degeneration of the nervous system of teenagers and young adults. If not treated, measles can progress into encephalomyelitis or pneumonia.

- Rhabdovirus. *Rhabdovirus,* such as the *rabies virus (Lyssavirus),* causes rabies following an animal bite. In rare cases, a person can be infected by inhaling the virus. Some animals such as bats pass the rabies virus through to their feces.

- Filoviridae. *Filovirus,* an example would be the *ebola virus* which causes African hemorrhagic fever.

Oncogenic viruses

Oncogenic viruses are viruses that produce tumors when they infect humans. The more common oncogenic viruses are:

- *Human papillomavirus (HPV).* Human papillomavirus causes common warts but also is believed to cause cervical cancer.
- Epstein-Barr virus (EBV). *Epstein-Barr virus* causes Burkitt's lymphoma, which is a tumor of the jaw. It is seen mainly in African children and causes a tumor in the nasopharyngeal (nasopharyngeal carcinoma).
- Herpes simplex virus 2 (HSV-2). *Herpes simplex virus 2* causes genital herpes, cervical cancer (cervical carcinoma), and oral lesions.
- Human T-cell leukemia virus 1 (HTLV-1). *Human T-cell leukemia virus 1* causes acute *T-cell lymphocytic leukemia,* which is a cancer that affects T-cell–forming tissues.
- Human T-cell leukemia virus 2 (HTLV-2). *Human T-cell leukemia virus 2* causes *atypical hairy cell leukemia.*

Plant viruses

Some viruses cause diseases in plants.
RNA plant virus:

- Picornaviridae: Includes the *bean mosaic virus.*
- Reovirus: Includes the *wound tumor virus.*

DNA plant virus:

- Papovaviridae: Includes the *cauliflower mosaic virus.*

Viroids

In 1971, plant pathologist O.T. Diener discovered an infectious RNA particle smaller than a virus that causes diseases in plants. He called it a viroid. Viroids

infect potatoes, causing potato, spindle tuber disease. They infect chrysanthe-
mums, stunting their growth. Viroids cause cucumber pale fruit disease. Millions
of dollars are lost each year in crop failures caused by viroids.

A viroid is similar to a virus in that it can reproduce only inside a host cell as
particles of RNA. However, it differs from a virus in that each RNA particle contains a single specific RNA. In addition, a viroid does not have a capsid or an
envelope. Some viruses do not have an envelope.

Prions

A *prion* is a small infectious particle that contains a protein. Some researchers
believe that a prion consists of proteins without nucleic acids because a prion is
too small to contain a nucleic acid and because a prion is not destroyed by agents
that digest nucleic acids.

Prion diseases referred to as transmissible spongiform encephalopathies
(TSEs) are progressive neurological diseases that are fatal to humans and animals. Researchers believe that prions cause Creutzfeldt-Jakob disease.
Creutzfeldt-Jakob disease is a neurological disease that causes progressive
dementia first observed by Hans Gerhard Creutzfeldt and Alfon Maria Jakob in
the 1920s. In 1976, Carlton Gajdusek won the Nobel prize for his work with the
TSE Kuru. Kuru is characterized by progressive ataxia incapacitation and death.

In 1982, neurobiologist Stanley Prusiner proposed that proteins cause the
neurological disease *Scrapie,* which is a degenerative neural conditon in sheep.
Prusiner named this infectious protein *prion*. Prions also cause other neurological diseases such as *Kuru* and *Gerstmann-Strausler-Sheinker syndrome*.

However, scientists are still studying prions to learn their origins and how prions replicate and cause disease.

Quiz

1. A virion is:
   (a) another name for a virus
   (b) a virus particle
   (c) a synthesized virus
   (d) a mature virus

2. A naked virus is
   (a) a virus without an envelope
   (b) a virus without a capsid
   (c) a virus without RNA
   (d) a virus without DNA

3. A bacteriophage is:
   (a) a virus that can be killed by antibiotics
   (b) a virus that acts like a bacteria
   (c) a bacteria that acts like a virus
   (d) a naked virus that uses bacteria as a host cell

4. What does a lytic virus inject into a host cell?
   (a) Nothing
   (b) Cytoplasm
   (c) Ribosomes
   (d) Nucleic acid

5. What is a capsid?
   (a) A capsid is the protein coat that encapsulates a virus
   (b) A capsid is the membrane bilayer of a virus
   (c) A capsid is another name for a bacteriophage
   (d) A capsid is the envelope around a virus

6. The envelope of a virus is made of:
   (a) pieces of the capsid
   (b) pieces of the host cell's membrane
   (c) pieces of the nuclei of the virus
   (d) pieces of the nuclei of the host cell

7. A state of lysogeny is:
   (a) when the envelope virus and the host cell interact with each other
   (b) when the envelope virus and the host cell don't interact with each other
   (c) when the envelope of a virus interacts with the capsid
   (d) when the envelope of a virus interacts with DNA or RNA

8. Oncogenic viruses
   (a) cause tumors
   (b) cause the common cold
   (c) cause herpes
   (d) cause influenza

9. What is a small infectious particle that contains a protein?
    (a) Viroid
    (b) Capsid
    (c) Virion
    (d) Prion

10. Genetic information of a virus is contained in:
    (a) the envelope
    (b) central nucleic acid
    (c) the prion
    (d) the viroid

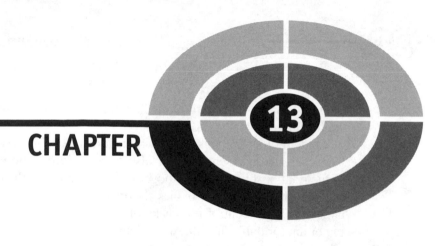

Epidemiology and Disease

No one wants to give a friend a cold, so we always cover a sneeze with our hands. However, the germs are still on our hands and are transferred to doorknobs, the food we touch, and to friends when we greet them with a handshake.

In this chapter, you'll learn about diseases and how disease are spread from one person to another. You'll also learn ways that outbreaks of disease are controlled and prevented by taking simple precautions.

What Is Epidemiology?

Epidemiology is the study of the distribution and determinants of diseases or conditions of a population. The word originated from the two Greek words *epi* ("among") and *demos* ("people"), or "among the people." An *epidemiologist* is a scientist trained to identify and prevent diseases in a given population. They are concerned with the *etiology,* or the specific cause of a disease in a given population. Epidemiologists use this information to design ways of preventing and controlling outbreaks of disease.

Epidemiology is considered a branch of microbiology because microorganisms cause many diseases. It can also be considered a branch of ecology because of the relationship among pathogens, their hosts, and the environment. The science of epidemiology provides the methods and information that are used to understand and control outbreaks of diseases in human populations, making it important for public health. An epidemiologist is a person who is trained to identify and prevent diseases in a given population or a medical doctor who is trained to identify and treat diseases in individual people.

Epidemiologists are concerned with the frequency or prevalence of diseases in a given population. An epidemiologist will identify the factors that cause disease or how that disease is transmitted and how the spread of communicable and noncommunicable diseases can be prevented. The *incidence rate* of a disease is the total number of new cases seen within a calendar year. The *prevalence* of a disease is the number of people infected at any given time. The *prevalence rate* is the total number of old and new cases of a disease. Frequencies are also expressed as proportions of the total population. The *morbidity rate* is the state of illness or the number of people in a given population that are ill. This is expressed as the number of cases per 100,000 people per year. The *mortality rate* is the number of people that are dead. This is measured as the number of deaths from a specific cause per 100,000 people per year.

Classification of Disease

Epidemiologists measure the frequency of diseases within a given population in regards to the geographical size of the area and the amount of damage the disease inflicts on the population. Diseases can be classified as endemic, sporadic, epidemic, or pandemic.

An *endemic disease* is the average or normal number of cases of a disease in a certain population. The number of people contracting the disease and the severity of the disease is so low that it raises little concern and does not constitute a health problem. An example is the varicella-zoster virus (the virus that causes chickenpox). Chickenpox is an endemic disease that usually affects children and is seasonal. An endemic disease can give rise to epidemics.

A *sporadic disease* occurs when there are small numbers of isolated cases reported. Sporadic diseases do not threaten the population.

An *epidemic disease* arises when the level of disease in a certain population exceeds the endemic level. This disease will cause an increase in mortality rate and the rate of morbidity. The level of destruction will be so large that it will cause a significant public health concern.

A disease becomes *pandemic* when it is distributed throughout the world. For example, in 1918 the swine flu (influenza) reached pandemic proportions. Some experts consider the HIV virus to be pandemic.

In a *common source epidemic*, large numbers of the population are suddenly infected from the same source. These epidemics usually are attributed to a contaminated supply of water or improperly prepared or handled food. An example is people who eat contaminated chicken salad at a college cafeteria. Everyone who eats the chicken salad on this particular day will become infected and feel ill. The epidemic will subside very fast, though, as the source of infection is eradicated.

A *propagated epidemic* occurs from person-to-person contact. The disease-causing agent moves from a person who is infected to a person who is not infected. In a propagated epidemic, the number of new cases rises and falls much slower than in common source epidemics, making the pathogen much harder to isolate and thus eliminate. An example is a "flu" virus.

Pathognomonic is a word that refers to the specific characteristics of disease. *Immunity* is the specific resistance to disease. *Virulence* is the degree of pathogenicity or the capacity of an organism to produce disease.

Pathology is the study of disease. It is derived from *Pathos* ("suffering") and *logos* which means ("science"). Pathology is that branch of discourse concerned with the structural and functional changes that occur due to a disease-causing agent or pathogen. A pathologist is a scientist or physician who studies the cause of diseases, or etiology, and pathogenesis, the manner in which a disease develops.

Infection Sites

The sites where a microorganism can infect a host organism are called *reservoirs of infection*. In these sites a microorganism can maintain its ability to cause infection. Reservoirs of infection include humans, some animals, certain non-living media, and inanimate objects.

HUMAN RESERVOIRS

Humans make good reservoirs because they can transmit organisms to other humans. Certain disease-causing agents have an *incubation period* during which they are contagious and can spread the disease even before a person exhibits signs or symptoms. These disease-causing agents can even be contagious during the recovery period.

When a person becomes *symptomatic* (feeling sick), the individual seeks medical attention and receives treatment. In many cases, however, diseases are spread from a person with subclinical findings—the symptoms are very mild and not recognizable. These individuals can spread the disease to a healthy person. *Asymptomatic* cases are a problem because infected people can infect other individuals without knowing they are infected. These individuals that are "carriers" of disease are called *disease carriers*.

Carriers of disease are classified as:

- *Subclinical carriers.* These individuals never develop clinical symptoms of the disease.
- *Incubatory carriers.* These individuals transmit the disease before becoming symptomatic.
- *Convalescent carriers* are individuals are recovering from the disease; however, they can still infect other people.
- *Chronic carriers.* These individuals develop chronic infections and transmit the infection for long periods of time.

ANIMAL RESERVOIRS

Many microorganisms can infect both humans and animals. Many of these disease-causing agents use animals as reservoirs of infection to infect humans. Apes and monkeys are good examples of animals that serve as reservoirs for human infection because they are physiologically similar to humans. When an animal infects humans, the humans can also serve as reservoirs for the infection.

A disease that is transmitted from domestic and wild animals to humans is called *zoonosis*. Two examples of zoonoses are anthrax, which is a bacterial disease that causes infection in dogs, cats, cattle, and other domestic animals. Humans become infected with direct contact with the animals, their wool, or hides, contaminated soil, inhalation of spores, and ingestion of meat or milk. Another is rabies, a virus that infects dogs, cats, skunks, wolves, and bats. Humans become infected through infected saliva in bite wounds. Humans and domestic animals can also be reservoirs for wild animals.

NONLIVING RESERVOIRS

Examples of nonliving reservoirs are water and soil. Soil is a good reservoir for the bacterium *Clostridium tetani*, which causes the disease tetanus. Contaminated water infected with human or animal fecal matter can contain many diseases-causing agents. An example is the bacterium *Vibrio cholera,* which causes Asiatic cholera, a disease caused by feces contaminated water where sanitation is poor. This organism invades the intestines and causes severe vomiting, diarrhea, abdominal pain, and dehydration.

Disease Transmission

A disease must have a *portal of exit* from the reservoir and a *portal of entry* to the infected host. This is how diseases are spread and new cases of infection occur. Examples of portal of exit are the respiratory tract, digestive tract, urinary tract, skin, and utero transmission. Diseases can spread by three different modes of transmission: contact transmission, vehicle transmission, and vector transmission.

CONTACT TRANSMISSION

Contact transmission of a disease-causing agent can either be *direct or indirect.* Direct contact transmission occurs from skin-to-skin contact, such as shaking hands, kissing, sexual contact, or making contact with open sores. Examples of diseases caused by contact transmission include herpes, syphilis, and staphylococcal infections.

Indirect contact transmission occurs when infection is spread through any nonliving, inanimate object. These contaminated inanimate objects are called fomites and include bedding, towels, clothing, dishes, utensils, glasses and cups, diapers, tissues, and even bars of soap. Examples of diseases caused by indirect contact transmission include the rhinovirus, hepatitis B, and tetanus.

Droplet transmission is a form of contact transmission that occurs through sneezing, coughing, and speaking in close contact with an infected individual. Examples of diseases transmitting in this manner are pneumonia, influenza, the common cold, and whooping cough.

VEHICLE TRANSMISSION

In vector transmissions, pathogens can be spread through the air and in water, food products, and body fluids (such as blood and semen). Airborne microorganisms mainly come from animals, plants, water, and soil. These microorgan-

isms can transmit disease through air. They can travel one meter or more through an air medium to spread infection.

Airborne pathogens have the greatest chance of infecting new individuals when these individuals are crowded together indoors or in a climate-controlled building where heating and air conditioning units regulate temperature and very little fresh air enters the building. Airborne pathogens can fall to the floor and combine with dust particles. This dust can then be stirred up with walking, dry mopping, or changing bedding and clothing. Examples of diseases that are transmitted by airborne transmissions and dust particles are measles, chickenpox, histoplasmosis, and tuberculosis.

Waterborne microorganisms that cause pathologies do not grow in pure water. They can survive in water with small amounts of nutrients but thrive in polluted water, such as water contaminated with fertilizer and sewage (which is rich in nutrients). Waterborne pathogens are usually transmitted in contaminated water supplies by either untreated or inadequately treated sewage. Indirect fecal-oral transmission of pathogens occurs when the disease-causing microorganism living in the fecal matter of one organism infects another organism. Bacterial pathogens infect the digestive system, causing gastrointestinal signs and symptoms. Examples of waterborne diseases are shigellosis and cholera.

Foodborne pathogens are normally transmitted through improperly cooked or improperly refrigerated food, or unsanitary conditions. Improper hygiene of the part of food handlers also plays a key role in foodborne transmission. Foodborne pathogens can produce gastrointestinal signs and symptoms. Examples of foodborne diseases are salmonellosis, typhoid fever, tapeworm, and listeriosos.

Vector Transmission

Vector spread is the transmission of an infectious agent by a living organism to humans. Most vectors are ticks, flies, and mosquitoes. These organisms are called *arthropods*. Vectors can transmit disease in two ways. First, mechanical vectors can passively transmit disease with their bodies. An example is the common housefly.

These animals commonly feed on fecal matter. They then fly to feed on human food, transmitting pathogens along the way. Keeping mechanical vectors away from food preparation and eating areas are means of prevention. Remember: The fly that is walking across your picnic lunch may have just walked across dog or cat feces. Examples of a few diseases transmitted by mechanical vectors are diarrhea caused by *E. coli* bacteria, conjunctivitis, and salmonellosis.

The second type of vectors are biological vectors and can actively transmit disease-causing pathogens that complete part of their life cycle within the vector. In most vector-transmitted diseases, a biological vector is the host for a phase of the life cycle of the pathogen. An example of a host organism is a mosquito that infects a human with malaria. Other diseases caused by biological transmission vectors are yellow fever, the plague, typhus, and Rocky Mountain spotted fever.

To cause infection, a microorganism must enter the body and have access to body tissues. The sites where microorganisms enter the body are called *portals of entry*. The portal of entry is similar to the portal of exit for the host to be susceptible for a certain disease. The portals of entry include the skin, digestive tract, respiratory tract, and urinary tract. Microorganisms can invade tissues directly or cross the placenta to infect the fetus. Skin that is intact prevents most microorganisms from entering the body, although some enter the ducts of sudoriferous glands (sweat glands) and hair follicles to gain entrance into the body.

Fungi can invade cells on the surface of the body and some can even invade other tissues. The larvae of parasitic worms can work their way through the skin and enter tissues. An example of a parasitic worm is the hookworm.

Mucous membranes make direct contact with the external environment. This allows microorganisms to enter the body. Examples of mucous membranes are the eyes, nose, mouth, urethra, vagina, and anus. The respiratory tract is an area of the body where microorganisms typically enter on dust particles that are inhaled with air or in acrosol droplets. Microorganisms that infect the digestive tract, are normally ingested with contaminated water or food, or even from biting the nails of contaminated fingers.

Many genitourinary infections are the result of sexual contact. Skin that is not intact due to injury, surgery, injections, burns, and bites makes it easy for invading microorganisms to penetrate body tissues. Common portals of entry are insect bites. Many parasitic diseases are caused by the bites of insects. Some diseases can affect the fetus through the placenta of an infected mother. Viruses such as the HIV virus, rubella (German measles), and the bacteria that cause syphilis behave in this way.

The transmission of disease by carriers causes epidemiological problems because carriers usually do not know they are infected and spread the disease, causing sudden outbreaks. Carriers can transmit disease by direct and indirect contact or through vehicles, such as water, air, and food.

The Development of Disease

In order for a pathogen to infect a host, there must be a susceptible host for the disease to be transmitted. If a host's *resistance* is low (resistance is the ability to ward off disease), its susceptibility increases (its chances of becoming infected increase). Primary defense mechanisms of the body for resistance include intact skin (no cuts or abrasions), mucous membranes, a good cough reflex, normal gastric juices, and normal bacterial flora.

If a microorganism penetrates these defenses the development of a disease process begins. First there may be an *incubation period*. This is the time between the initial exposure and start of the infection to the first appearance of signs and the feeling of symptoms. Different microorganisms have different incubation periods. An examples is the Epstein-Barr virus, which causes infectious mononucleosis and has an incubation period of two to six weeks.

The varicella-zoster virus, which causes varicella (chickenpox), has an incubation period of two weeks. The human immunodeficiency virus, the virus that causes AIDS, has an incubation period of 7 to 11 years. During this phase, the disease can be spread from the infected individual to a non-infected individual.

The *prodromal period* follows the incubation period. This period presents with mild symptoms.

The *period of illness* is the acute phase of the disease. Here the individual presents with signs and symptoms of the disease. *Signs* are objective findings that an observer or physician can see. These are physical changes that can be measured. Examples of signs are fever, skin color or lesions, blood pressure, inflammation, and paralysis.

Symptoms are subjective and cannot be seen by an observer. Symptoms present as changes in bodily functions, such as pain, numbness, chills, general fatigue, or gastrointestinal discomfort. It is in this period of the disease where white blood cells may increase and the individual's immune system responds to combat the disease-causing pathogen. If the individual's defense mechanism of the immune system does not successfully overcome the disease or if the disease is not treated properly, the person can die.

During the *period of decline*, the individual's signs and symptoms subside and the person feels better. This period may take 24 hours to several days. During this time, the individual is prone to secondary infections.

The *period of convalescence* is the phase where recovery has occurred. The body regains strength and is returned to a state of normality. During this phase, infection can also be spread.

Epidemiological Studies

Epidemiological studies began in 1855 with the work of English physician John Snow. Snow conducted studies relating to the cholera outbreak in London, England. Snow, through careful analysis of deaths related to cholera, case histories of victims, and interviews with survivors, traced the source of the epidemic to a water pump. He concluded that individuals who died from cholera drank water contaminated with human feces. Snow's study and his method of analyzing where and when a disease could occur and the transmission of that disease within a given population gave way to a new approach in medical research and epidemiological studies.

It was not until 1883 that the cholera bacterium *Vibrio cholerae* was identified by Robert Koch. After the studies of Snow, other investigators conducted epidemiological studies. Epidemiologists now use three types of studies when determining the occurrence of disease: descriptive, analytical, and experimental.

Descriptive epidemiology is the collection of all data described in the occurrence of disease. This data includes the number of cases, which portion of the population was infected, where the cases occurred, and information about the affected individuals (race, sex, age, occupation, marital status, and socio-economic status).

Analytical epidemiology analyzes the cause of a disease and the effect of the disease in a given population. Here epidemiologists compare a group of people who have the disease with people who do not have the disease.

Experimental epidemiology are studies designed to test hypotheses for the source of a particular disease. Experiments are conducted on human or animal subjects to test the hypothesis. An example is the testing of an experimental drug designed to control a specific disease. A group of infected individuals is divided randomly so that one group receives the drug and the other receives a placebo. A *placebo* is a substance that has no effect on the individual receiving it, but the individual believes he or she is receiving treatment. If those people who were treated with the drug recover faster than those taking the placebo, investigators can conclude that the drug treatment is effective.

Control of Communicable Diseases

There are different ways of controlling or limiting the spread of communicable diseases. These methods include isolation, quarantine, immunization, and vector control.

Isolation requires that a patient infected with a communicable disease be prevented from making contact with the general public. The Centers for Disease Control and Prevention has designated five categories of isolation: strict, protective, respiratory, enteric, and wound and skin.

Quarantine requires the separation of animals and humans that have been infected or exposed to a communicable disease from the general public.

Immunization is an effective means of controlling the spread of communicable diseases by the use of safe vaccines. A vaccine is a preparation of killed, attenuated, inactivated, or fully virulent organisms that are administered to induce or produce artificially acquired active immunity.

Vector control is a good way of controlling the spread of infectious disease when the vector, such as rodents or insects, is identified. This vector's habitats and breeding grounds can be treated with insecticides and poisons. Also, barriers such as window screens, netting, and repellents can provide protection against bites and infection.

Nosocomial Infections

A nosocomial infection is an infection that is the result of a pathogen that was acquired in a hospital or clinical care facility. Nosocomial is derived from the Latin word *nosocomium*, which means "hospital." These are the diseases that a patient can obtain when he or she is being cared for in a hopital. These diseases can also affect the caregivers, such as the hospital staff, nurses, doctors, aides, and even visitors or anyone else who has contact with a hospital or medical facility.

The Centers for Disease Control and Prevention estimate that among the patients that are admitted to hospitals, 5 to 15 percent acquire some type of nosocomial infection. Nosocomial infections result directly in over 20,000 deaths and indirectly in about 60,000 deaths.

Nosocomial disease-causing pathogens come from either endogenous or exogenous sources. *Endogenous* infections are caused by pathogens that were brought into the hospital by the patient; the opportunistic pathogen is among the patient's own microbiota.

Exogenous infections are caused by organisms that enter the patient's body from the external environment. These organisms can be acquired from animate sources, such as hospital staff, other patients, or people visiting the hospital. Organisms can also come from inanimate sources, like hospital equipment, intravenous and respiratory therapy equipment, catheters, computer keyboards, bathroom fixtures, doorknobs, soaps, and even certain disinfectants.

Who Is Susceptible?

A person who is susceptible to a nosocomial infection is one who has a compromised immune system. Compared to the general public, patients in hospitals have a lowered resistance to disease, thus making them more susceptible to infections. Many of these patients are said to be "compromised hosts"—they have breaks in their skin due to accidental or surgical wounds (lesions, bedsores, or burns). Others may have compromised mucous membranes that line the respiratory tract, the digestive tract, or the urinary and reproductive system, making them more susceptible to disease-causing pathogens.

Prevention and Control of Nosocomial Infections

Most hospitals and clinical facilities have control measures and procedures aimed at preventing nosocomial infections. For a hospital to be accredited, it must have a designated person responsible for developing and implementing policies and procedures that would control infections and communicable diseases. This person can be a registered nurse or an epidemiologist.

PUBLIC HEALTH ORGANIZATIONS: THE CENTERS FOR DISEASE CONTROL

Public health agencies have been created in cities, countries, states, and at the federal level. The United States has recognized the importance of identifying and controlling infectious diseases since the eighteenth century. The Centers for Disease Control and Prevention (CDC) is a branch of the United States Public Health Service (USPHS). The CDC is located in Atlanta, Georgia, and is the central source of epidemiological study and information. The CDC is responsible for the control and prevention of disease, public education, and occupational health and safety.

Quiz

1. The number of people infected by a disease at any point in time is called
   (a) incidence
   (b) prevalence
   (c) morbidity rate
   (d) prevalence rate

2. The total number of old and new cases of a disease is called
   (a) incidence
   (b) prevalence
   (c) morbidity rate
   (d) prevalence rate

3. The number of people in a given population who are ill is called
   (a) incidence
   (b) prevalence
   (c) morbidity rate
   (d) prevalence rate

4. What is a disease called when there is a worldwide distribution of the disease?
   (a) Pandemic
   (b) Sporadic disease
   (c) Source epidemic
   (d) Propagated epidemic

5. What is it called when a disease arises from person-to-person contact?
   (a) Pandemic
   (b) Sporadic disease
   (c) Source epidemic
   (d) Propagated epidemic

6. What is it called when large numbers of a population are infected suddenly from a common source?
   (a) Pandemic
   (b) Sporadic disease
   (c) Source epidemic
   (d) Propagated epidemic

7. What is it called when there are small isolated occurrences of a disease reported?
   (a) Pandemic
   (b) Sporadic disease
   (c) Source epidemic
   (d) Propagated epidemic

8. What is the period of time called when a disease is contagious and communicable before showing signs and symptoms and during recovery?
   (a) Symptomatic
   (b) Incubation period
   (c) Asymptomatic
   (d) Carrier period

9. Incubatory carriers are individuals who transmit the disease before becoming symptomatic.
   (a) True
   (b) False

10. Convalescent carriers are individuals who develop chronic infections and transmit the infection for long periods of time.
   (a) True
   (b) False

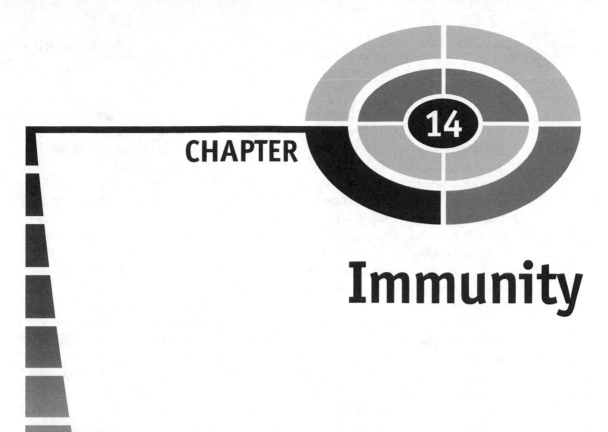

CHAPTER 14

Immunity

There's a battle going on beneath your skin. Thousands of microbes crawl into every nook and cranny of your body in a daily invasion. Yet you rarely notice because an army of B cells, T cells, natural killer cells, and other parts of your immune system counterattack, ripping most microbes to pieces. Those that survive give you a runny nose, cough, and feverish feeling, and, on rare occasions, more serious illness.

The immune system fights invading bacteria, viruses, and pollen and venom from bees. Anything foreign that enters your body is surrounded and either neutralized or destroyed by your immune system before it can cause damage to your body.

In this chapter, you'll learn the parts of the immune system and how each functions to give you protection days and years from now.

What Is Immunity?

An *immune system* is an organism's protection from invading organisms and foreign substances, such as bacteria, viruses, fungi, helminths, protozoa, pollen, transplanted tissues, and insect venom. Parts of these substances are called *anti-gens,* or immunogens. Antigens can be polysaccharides or proteins and provoke a specific immune response in organisms. *Immunity* is a specific defensive response of a host when a foreign substance or organism invades it.

The body reacts to the foreign substance (antigen) by forming antibodies. Antibodies are proteins that are made by the body in response to an antigen and can combine specifically with that antigen.

The immune system recognizes a body or substance within the organism as self or nonself. Self is any body or substance that belongs to the organism. Non-self is any body or organism that doesn't belong to the organism. An antigen is recognized as nonself by the immune system.

An antigen causes the organism to form *antibodies* and *specialized lympho-cytes* that target the specific antigen. If the antigen invades again, these antibod-ies and specialized lymphocytes attack the antigen, making the antigen inactive or destroyed. This is called an *immune response* against an antigen.

Cells that become cancerous are recognized as foreign and can be destroyed. Cancerous cells, once established as a tumor, might have a dreadful effect on the organism because the immune system is not be able to fight them.

ACQUIRED IMMUNITY

The development of antibodies and specialized lymphocytes is called *acquired immunity* because it is acquired over an organism's life through naturally acquired immunity and artificially acquired immunity means.

Acquired immunity is the protective defense mechanism an organism devel-ops against foreign substances and microorganisms. This type of immunity is established throughout an individual's life.

There are two types of acquired immunity. These are active and passive. When an individual is exposed to a disease-causing microorganism or foreign substance, the person's own immune system responds by making its own anti-bodies and lymphocytes.

In *passive immunity,* already-made antibodies are introduced or passed on to an individual. The individual does not make its own antibodies.

Naturally acquired active immunity occurs when an individual is exposed to an infectious disease. The individual's immune system responds by making its own antibodies and lymphocytes (T cells and B cells) on its own.

Naturally acquired passive immunity occurs when antibodies (IgG) are made by a mother and passed on to the fetus through the placenta. IgA antibodies are also passed to the baby in the first secretion of breast milk, called *colostrum,* during breast-feeding.

Artificially acquired active immunity occurs when an individual is given a vaccine. A vaccine is a substance that contains the weakened or dead organisms. These antigens stimulate the immune response, but do not cause major sickness. The body remembers the antigen with memory cells the next time there is exposure to the antigen.

Artificially acquired passive immunity occurs when antibodies are developed outside the individual and intravenously injected into the body. This form of immunity helps the body's own defenses in combating infection.

Serum and Antibodies

Serum is the fluid that remains after the blood has clotted and the serum is separated from the blood cells and clotted material. Electrophoresis is a laboratory technique that introduces electrical current into a serum placed in a gel. This causes proteins within the serum to move across the gel at different rates, which represents different globulins.

There are three globulins. These are called *alpha*, *beta*, and *gamma* globulins. *Gamma globulin* contains the most antibodies. A serum rich in antibodies is called either *gamma globulin* or *serum globulin*.

Gamma globulin can be taken from a person who is immune to an antigen and injected into a person who lacks the antigen, who then immediately becomes immune from the antigen. However, the immunity lasts for about three weeks, at which time the antibodies degrade.

Types of Immunity

ANTIBODY-MEDIATED IMMUNITY

Antibody-mediated immunity, which is also known as *humoral immunity*, uses antibodies in extracellular fluids, such as mucus secretions, blood plasma, and lymph, to combat antigens. These antibodies, produced from *B cells*, which are also known as *B lymphocytes*, primarily attack bacteria, bacterial toxins, and viruses that invade body fluids. They also attack transplanted tissues.

Antibody-mediated immunity was discovered by German scientist Emil von Behring at the turn of the twentieth century in his quest to create an immunization against diphtheria. Behring called this humoral immunity because the medical community called body fluids *humors*.

CELL-MEDIATED IMMUNITY

Cell-mediated immunity involves specialized lymphocytes called T cells, also known as *T lymphocytes*, to attack foreign organisms rather than using antibodies. T cells are also effective against helminths, fungi, and protozoa. In addition, T cells regulate aspects of the immune system.

Cell-mediated immunity was explained by Russian biologist, Elie Metchnikoff, who in the early 1900's noticed that phagocytic cells were much more effective in animals that were immunized. This immunity was used in the mid-twentieth century to protect people against tuberculosis.

A Closer Look at Antigens

An antibody identifies its corresponding antigen by one or more regions on the antigen known as the *epitopes*, which are also called antigenic determinants. The epitope must be the right size, shape, and chemical structure for the antibody to bind to the epitope and then proceed to disable or destroy the antigen.

Antigens tend to have a molecular weight of 10,000 or more yet some foreign substances might have a lower molecular weight and are not antigens. They are called *haptens* and must attach themselves to a large carrier molecule in order to become antigenic. Antibodies only attack the hapten and not the carrier molecule.

Penicillin is a common hapten. Penicillin does not have an antigen effect in most people. However, when penicillin attaches to serum proteins an allergic reaction results in some people. These people are said to be allergic to penicillin. An allergic reaction is a typical immune response.

Antigens can be proteins, large polysaccharides, lipids, or nucleic acids. However, antigens that are lipids and nucleic acids must be combined with proteins and polysaccharides; otherwise they are not antigens.

A Closer Look at Binding

Antibodies are known as *immunoglobulins* (Igs), which is a group of soluble proteins. An antigen can cause the production of different antibodies if the antigen has several epitopes. Epitopes or *antigenic determinants,* are known as *antigen-*

binding sites. The number of antigen-binding sites is called the antibody's *valence*. There are at least two antigen-binding sites on each antigen where a human antibody can bind. This is referred to as a *bivalent antibody* because the antibody's valence value is two.

The structure of a bivalent antibody is called a *monomer* and consists of four protein chains that are named after their relative molecular weights. These are two *light (L) chains* and two *heavy (H) chains*.

These protein chains are joined together to form a *Y*-shaped molecule that is flexible enough to form a *T*-shaped molecule. There are two regions of the protein chain. These are the variable (V) region and the constant (C) region.

The *variable region* is located at the ends of the arms of the *Y*. These are the sites where the antibody binds with an antigen. The variable region is a three-dimensional structure of amino acid sequences whose structure reflects the epitopes of the antigen.

The *constant region* is the stem of the *Y* and is called the *Fc region*. The Fc region binds adjacent complementary antibodies if both antigen-binding sites are attached to an antigen; otherwise the adjacent complementary antibodies are free to attach and react with antigen.

There are five types of constant regions each associated with the five classes of immunoglobulin.

IgG

IgG immunoglobulin neutralizes bacterial toxins and attacks circulating bacteria and viruses by enhancing the effectiveness of phyagocytic cells. Nearly 80 percent of the antibodies in serum are IgG. IgG can cross blood vessel walls and the placenta and can enter tissue fluids.

IgM

IgM immunoglobulin is the first antibody to respond to an antigen or initial infection and makes up about 10 percent of antibodies in serum. IgM is relatively large in size and has a pentamer structure of five monomers bonded together by a joining (J) chain. This chain is a polypeptide. Because of its size, IgM remains in blood vessels and not in tissue fluids. IgM responds to the ABO blood group antigens and enhances the effectiveness of phyagocytic cells. When initial exposure to an antigen occurs, IgM antibodies are the first to appear.

IgA

IgA immunoglobulin is the most common antibody in body secretions and in mucous membranes. It consists of about 15 percent of the antibodies in a serum. IgA protects infants from gastrointestinal infections and fights antigens that affect the respiratory tract. Plasma cells in the mucous membrane form secretory IgA, which is then passed through the mucosal cell and attacks antigens on the mucosal surface such as bacteria and viruses. IgA is short-lived

IgD

IgD immunoglobulin is found in blood and lymph fluid and is an antigen receptor on the surface of B cells. This also provides protection against parasitic worms.

IgE

IgE immunoglobulin binds to basophil cells and mast cells that release chemical mediators, such as histamine, that cause an allergic reaction. When pollen is the antigen, the allergic reaction is referred to as hay fever. IgE is less than one percent of serum antibodies. This provides protection against parasitic worms.

B Cells

B cells are cells that develop from stem cells in the bone marrow and the liver of fetuses. They are transported to the lymph nodes and spleen where they use *antigen receptors*, also known as *antigen-binding sites*, on the cell's surface to seek out antigens.

Once an antigen is detected, the B cell with T cells activates a special group of lymphocytes that produces antibodies used in the antibody-mediated immunity response. T cells do not make antibodies. When B cells come in contact with extracellular antigens, the B cell transforms into *plasma cells*, that produce antibodies at about 2,000 antibodies per second to combat that antigen.

Memory cells are also produced when a B cell is stimulated by an antigen. A memory cell provides the organism with long-term immunity to the antigen.

B cells react to one kind of antigen that is referred to as its *complementary antigen* and are able to identify that antigen because antigen receptors bind to one specific antigen. Here's how it works: Once the antigen binds to the antigen receptor, the B cell replicates into a clonal selection. A *clonal* selection is a large cluster of clone cells.

An antibody attaches to an antigen at an antigen-binding site to form an *antigen-antibody complex*. This complex is very specific. However, when there are large quantities of antigens, the antigens attach to antibodies where they do not exactly fit. This makes for less-than-perfect matches for the antigen-antibody complex. These antibodies are said to have less affinity to the antigen.

B cells under go apoptosis if the B cell does not come in contact with an antigen. *Apoptosis* is a programmed death of the B cell that causes phagocytes to remove the cell from the organism.

STRATEGIES FOR COMBATING ANTIGENS

The formation of antigen-antibody complexes is useful in the response to infectious organisms or foreign substances because they remove the infectious agent from the body. Protective methodology of binding antibodies to antigens is accomplished by agglutination, opsonization, neutralization, antibody-dependent cell-mediated cytotoxicity, and the activation of complement.

- *Agglutination* generates antibodies that clump together antigens, making them easy to ingest by phagocytes.

- *Opsonization* coats the antigen with antibodies to make it easy for phagocytic cells to ingest and for lysis.

- *Neutralization* blocks antigens from attaching to targeted cells, thereby neutralizing the antigen.

- *Antibody-dependent cell-mediated cytotoxicity* coats the foreign cell with antibodies. Nonspecific immune cells then destroy the foreign cell from the outside. This is used for organisms that are too large for phagocytic cells to ingest.

- *The activation of complement* is used when infectious agents are coated with reactive proteins that cause IgG and IgM antibodies to attach to the agent, causing *lysis* of the cell membrane and resulting in ingestion by phagocytes.

Lasting Immunity

When antigens are first encountered, the *primary immune response* occurs causing an increase in the antibody titer. The *antibody titer* is the amount of antibodies in serum of the infected organism. There are no or undetectable levels of antibodies when the antigen first attacks the organism. Then the antibody titer gradually increases and then declines as the antigen is destroyed or neutralized.

When antigens are encountered for the second time, the *secondary immune response* occurs, causing memory cells to quickly transform into plasma cells that produce antibodies. The secondary immune response is also known as the *anamnestic response* or *memory response*. A *memory cell* is a B lymphocyte that was generated in the primary immune response but did not became an antibody-producing plasma cell at that time.

Antibodies Used for Diagnosing Diseases

Antibodies are useful in diagnosing diseases since a particular antibody is produced only if a complimentary antigen is present in the organism. This is useful in identifying an unknown disease-causing pathogen.

Antibodies can be produced in the laboratory by a clone of cultured cells that make one type of antibody. These antibodies are called monoclonal antibodies. Malignant cells of the immune system called myeloma cells are used because they divide forever. This is why malignant cells (cancer cells) are so devastating to our body. These myeloma cells are then mixed with lymphocytes that have been designed to produce a specific antibody. When these cells are mixed together, they fuse to become one cell called a *hybridoma*. Hybridomas divide indefinitely because they have the gene from the myeloma cell. They can produce large amounts of antibodies because they have the gene from the lymphocyte.

Monoclonal antibodies are used to diagnose streptococcal bacteria and chlamydial infections. Some over the counter pregnancy tests use monoclonal antibodies to detect the hormones found in urine during pregnancy.

Chemical Messengers

Cells in the immune system communicate with each other by using chemical messengers sending signals to trigger activities. These chemical messengers are known as *cytokines*. There are 60 known cytokines.

Table 14-1.   Important Interleukins

Interleukins	Description
Interleukin-1	Stimulates T cells Attracts phagocytes in an inflammatory response
Interleukin-2	Stimulates B cell production Stimulates T cell production
Interleukin-8	Attracts phagocytes to the inflammation site
Interleukin-12	Stimulates the differentiation of CD4-type T cells

Cytokines used for communication between leukocytic cells are called *interleukins*. There are 17 known interleukins that are identified by numbers assigned to them by an international committee. Table 14-1 lists important interleukins.

Cytokines are used as therapeutic agents to combat disease. For example, Interleukin-1 is used to stop blood flow to tumors in animals, thereby killing the tumor.

T Cells

T cells develop from stem cells in bone marrow and migrate to the thymus gland where they mature. They then migrate to the lymphatic system to begin their fight against antigens. The "T" stands for thymus gland. Once the organism reaches late adulthood, the ability to create new T cells diminishes, resulting in a weaker immune system as the organism ages.

A T cell attacks a specific antigen that is displayed on the surface of the cell. These cells are called antigen-presenting cells (APCs), such as, macrophages and dendritic cells. After the antigen is ingested by the APC, fragments of the antigen are placed on the surface of the cell. These fragments must be near the cell-surface self-molecules. Self-molecules are part of the histocompatibility complex (MHC), which is a group of proteins that are unique to a person and used to distinguish self from nonself.

When an antigen receptor encounters fragments of the complimentary antigen, the T cell transforms into the *effector T cell* that carries out the immune response. An effector T cell is an *antigen-stimulated cell*. Some T cells attack

the antigen in a primary immune response, while others become memory cells and take on a secondary immune response role when the antigen is encountered later on.

There are four types of T cells, each identified by characteristics of their surface molecules. These are

- *Helper T (T_H) cells.* These cause the formation of cytotoxic T cells, activate macrophages, produce cytokines, and are essential to the formation of antibodies by B cells.
- *Cytotoxic T (T_C) cells.* These destroy cells that have been infected by viruses and bacteria.
- *Delayed hypersensitivity T (T_D) cells.* These are associated with allergic reactions.
- *Suppressor T (T_S) cells.* These turn off the immune response when there are no antigens.

T cells are also identified by their surface receptors, called clusters of differentiation (CD). There are two types of clusters of differentiation. These are:

- CD4—Helper T cells
- CD8—Cytotoxic T cells and suppressor T cells

Macrophages and Natural Killer Cells

Macrophages are phagocytic cells that ingest antigens. They are in a resting state until they receive a cytokine from the helper T cell, at which point they become large and ruffled and ready to attack the antigens. Macrophages destroy virus-infected cells and bacteria in intracellular locations. They also eliminate some cancerous cells.

Natural killer cells (NK) are lymphocytes that destroy other cells such as tumor cells. Natural killer cells are always active and searching for an infected cell. These are different from other cells in the immune system, which become activated only when stimulated by an antigen.

Quiz

1. What triggers the immune response when a foreign substance invades the body?
   (a) Acquired immunity
   (b) Antigen
   (c) Antibody
   (d) Self

2. Immunity where a mother passes antibodies to her fetus is called
   (a) passive naturally acquired immunity
   (b) active naturally acquired immunity
   (c) passive artificially acquired immunity
   (d) active artificially acquired immunity

3. Immunity where antibodies are developed outside the organism in an immune serum is called
   (a) passive naturally acquired immunity
   (b) active naturally acquired immunity
   (c) passive artificially acquired immunity
   (d) active artificially acquired immunity

4. Immunity where the organism produces antibodies and specialized lymphocytes to fight an antigen is called
   (a) passive naturally acquired immunity
   (b) active naturally acquired immunity
   (c) passive artificially acquired immunity
   (d) active artificially acquired immunity

5. Immunity where antigens are injected into an organism as a vaccine is called
   (a) passive naturally acquired immunity
   (b) active naturally acquired immunity
   (c) passive artificially acquired immunity
   (d) active artificially acquired immunity

6. A serum that contains most antibodies is called
   (a) alpha globulin
   (b) gamma globulin

(c) beta globulin

(d) antibody degrade

7. What immunity uses antibodies in extracellular fluids?
    (a) Antibody-mediated immunity
    (b) Cell-mediated immunity
    (c) Epitopes
    (d) Haptens

8. What kind of immunity uses specialized lymphocyte cells called T cells?
    (a) Antibody-mediated immunity
    (b) Cell-mediated immunity
    (c) Epitopes
    (d) Haptens

9. The number of antigen-binding sites is called the antibody's valence.
    (a) True
    (b) False

10. The structure of a bivalent antibody is called a monomer
    (a) True
    (b) False

CHAPTER 15

Vaccines and Diagnosing Diseases

Would you pay your doctor to inject you with the flu virus? Of course not, but that's what millions of people do each year—and you might be one of them if you get an annual flu vaccine. The flu vaccine contains some form of the flu virus or an attenuated form.

Vaccines prevent you from catching a disease because they contain an element of the disease. This element triggers your body to create an army of antibodies that are trained to stamp out the disease if you contract it in the future.

There are many kinds of vaccines, each designed to prevent certain diseases. You'll learn about vaccines in this chapter. You'll also learn how your immune system is used to diagnose diseases.

What Is a Vaccine?

A *vaccine* is a suspension that contains a part of a pathogen that induces the immune system to produce antibodies that combat the antigen. The concept of a vaccine stems from the variolation process that was used in eighteenth-century England to protect people from smallpox.

The *variolation process* requires that a needle tip of smallpox be placed in the vein of a patient. Nearly all the patients contracted a mild case of smallpox, which left them with antibodies that protected them from contracting the disease. Half of the patients who contracted smallpox died. By contrast, only one percent who received the variolation process died.

Edward Jenner noticed that dairymaids who contracted cowpox, which is related chemically to smallpox, were immune to smallpox. Jenner discovered that injecting cowpox into the skin of a healthy person prevented them from developing smallpox. Jenner's discovery enabled Louis Pasteur to develop the technique of creating vaccines.

The injection of an antigen induces the primary immune and secondary immune responses in the patient. The primary immune response produces antibodies and the secondary immune response produces memory cells that attack a future invasion of the antigen (see Chapter 14).

Vaccines play an important role in controlling the spread of viruses. A virus cannot be treated with antibiotics. However, you can minimize catching the flu by getting a flu shot, which is a vaccine against a particular strain of flu virus.

Vaccines also prevent bacterial infections such as typhoid, but are not as effective on bacteria as they are on viruses. However, bacteria infections are treatable with antibiotics, which is a common method of combating bacterial diseases.

Scientists use vaccines to provide herd immunity to a population. *Herd immunity* requires that most—not all—of the population be immunized in order to prevent an epidemic of a disease. An outbreak of a disease would be isolated to a small percentage of the population and therefore have a minimum effect.

Types of Vaccines

There are six types of vaccines. These are:

Attenuated whole agent. These vaccines are designed for people who have a normal immune system. The attenuated whole agent vaccine uses weakened living microbes to mimic the real infection to produce 95 percent immunity over a long term without the need of a supplemental vaccination called a *booster*. Common attenuated whole agent vaccines include those for tuberculosis bacillus, measles, rubella, Sabin polio, and mumps. There is a risk that live microorganisms or virus can regain their strength resulting in the patient contracting the disease.

Inactivated whole agent. These vaccines are not designed for people who have an abnormal immune system. The inactivated whole agent vaccine uses dead microbes that were killed by phenol or formalin. Common inactivated whole agent vaccines include those for pneumonia, Salk polio, rabies, influenza, typhoid, and pertussis (commonly known as whooping cough).

Toxoids. The toxoid vaccine is made of toxins produced by a virus or bacteria that has been inactive. They are then used against toxins that are produced by a disease-causing microorganism. Patients require a booster vaccination every 10 years because the toxoid vaccine does not provide lifelong immunity. Common toxoid vaccines include those for diphtheria and tetanus.

Subunit. These vaccines have few side effects. The subunit vaccine uses fragments of a microorganism to create an immune response. Subunit vaccines are produced by using genetic engineering techniques to insert the genes of an antigen into another organism are called *recombinant vaccines.* Common subunit vaccines include those for hepatitis B.

Conjugated. These are fairly new in development and are designed for children under 24 months whose immune system normally does not respond well to vaccines based on capsular polysaccharides. These polysaccharides are T-independent antigens. The vaccine is produced by combing the polysaccharides with a protein.

Nucleic acid. These vaccines are in the animal testing stage. The nucleic acid vaccine, which is also called the *DNA vaccine*, contains plasmids of naked DNA and is designed to produce protein that stimulates an immune response. The nucleic acid vaccine has a strong effect on large parasites and viruses.

Developing a Vaccine

Vaccines are developed by cultivating a large quantity of pathogen, which is a disease causing organism. Some pathogens, such as the rabies virus, can be cultivated in animals. For example, a chick embryo is commonly used to grow viruses and is the method used to develop the influenza vaccine.

When vaccines were first introduced against measles and polio they would only grow in humans. However, with the development of cell culture techniques, cells from humans and primates enable large-scale viral growth. Scientists use recombinant vaccines because they do not need an animal host to grow the microorganism. An example is hepatitis B.

Diagnosing Diseases

The reaction between an antibody and complementary antigen is used in conjunction with observing symptoms to diagnosis a disease. You have probably seen this diagnosis technique used when you were tested for tuberculosis. This test required that a suspension of *Mycobacterium tuberculosis* be injected into your skin. If the site becomes red in a couple of days, then you tested positive for tuberculosis. The redness is a reaction between antibodies and antigen.

Scientist use eight types of reactions to diagnose diseases. Each determines if a specific antibody or a specific antigen is present based on its complementary antibody or antigen. These are:

Precipitation reaction. IgG or IgM antibodies are combined with soluble antigens. If the antibody and antigen are in an optional ratio, they form an antigen-antibody complex called a lattice. The reaction occurs immediately, however the lattice may not form until minutes after the reaction begins. There are three commonly used precipitation reaction tests. These are:

- *Precipitin ring test.* A ring appears in the area of the optimal ratio, which is called the zone of equivalence.

- *Immunodiffusion test.* A line is visible in the area of the optimal ratio.

- *Immunoelectrophoresis test.* Uses electrophoresis to identify separated proteins in human serum. The test is called the Western blot test and is used in AIDS testing.

Agglutination reaction. Particulate antigen or soluble antigens that adhere particles are introduced to antibodies to form an aggregate reaction called *agglutination* (clumping of cells). There are two types of agglutination reaction tests. These are:

- *Direct agglutination test.* Uses a plastic microtiter plate to detect antibody reaction with large cellular antigens. The microtiter plate contains a series of wells. Each well has the same amount of antigen and successively diluted serum of the antibody. The direct agglutination test measures the titer antibodies in the serum. The titer is lower at the onset of the disease and higher later in the disease.

- *Indirect agglutination test.* Soluble antigens are absorbed onto the surface of latex spheres where they react to antibodies. This has a relatively short reaction time, enabling a diagnosis in 10 minutes. Indirect agglutination test is commonly used to diagnosis streptococci, which cause sore throats.

Hemagglutination reaction. This type of reaction uses red blood cell surface antigens and complimentary antibodies to determine if red blood cells clump in reaction to the antibodies. This test is used in typing blood and in diagnosing infectious mononucleosis. If the antigen is a virus the reaction is called *viral hemagglutination.*

Neutralization reaction. This reaction uses antibodies as an antitoxin to block the extoxin or toxoid of bacteria or a virus. *Extoxin* is an active toxin and *toxoid* is an inactive toxin. A serum containing the antibody for a particular antigen is placed in a cell culture that contains cells and the antigen. If the antigen does not destroy (*cytopathic effect*) cells, then the extoxin is neutralized. Therefore, the antigen is complementary to the antibody. Tests that use the neutralization reaction are called *vitro neutralization tests*. One such test is the *hemagglutination inhibition test*, which is used to diagnose a number of infections including measles, influenza, and mumps.

Complement-fixation reaction. This reaction uses a group of serum proteins called a *complement* to bind with the antigen-antibody complex. The serum is said to be fixed when the serum total binds to the antigen-antibody complex.

Fluorescent-antibody (FA) reaction. This reaction combines fluorescent dyes with an antibody, making the antibody fluorescent when exposed to ultraviolet light, and is used to detect a specific antibody in a serum. The fluorescent-antibody reaction is used to test for rabies. There are two types of fluorescent-antibody reaction tests. These are:

- *Direct FA test.* The antigen and the fluorescien-labeled antibodies are combined on a slide and then incubated. The slide is washed to remove any antibodies that did not attach to the antigen and then observed under a fluorescence microscope for yellow-green fluorescence.

- *Indirect FA test.* The antigen and the serum containing the antibody are combined on a slide. Fluorescein-labeled antihuman immune serum globulin (antiHISG) is also added to the slide, which reacts to human antibody on the slide. The slide is incubated and washed, then observed under a fluorescence microscope. If the antigen is fluorescent, then the complementary antigen is on the slide.

Enzyme-linked immunosorbent assay (ELISA). This is an objective test for either antigens or antibodies using a reagent. There are two types of test. These are:

- *Direct ELISA.* The objective is to identify the antibody for a specific antigen. The antibody is placed in the microtiter plate wells where it adsorbs to the surface of the well. The antigen is then placed in each well. Wells are then washed. The antigen that reacts to the antibody will be retained and remain there after the washing. Next a second antibody for the antigen is added and adsorbed to the well wall. The antigen is sandwiched between two antibody molecules if the test is positive; otherwise the test is negative. Direct ELISA is used to detect drugs in urine.

- *Indirect ELISA.* The objective is to identify the antigen for a specific antibody. A known antigen is placed in the microtiter plate wells where it adsorbs to the surface of the well. The antibody is placed in each well and will bind to the antigen to form an antigen-antibody complex. The wells are washed to remove antibodies that did not bind. AntiHISG that has been linked with an enzyme is then placed into the well and reacts to the antigen-antibody complex in the well. All unbound HISG is rinsed. A substrate for the enzyme is added. If there is a color change, then the test is positive.

Radioimmunoassay (RIA). Uses radioactive tags to mark antigens and antibodies. Samples are then scanned to determine the presence of the tag.

Quiz

1. What process requires that a needle tip of antigen be placed in a vein of a patient to give the patient immunity to the antigen?
    (a) Half-life process
    (b) Jenner process
    (c) Variolation process
    (d) Antiviral process

2. Herd immunity requires an entire population to be immunized against an antigen.
    (a) True
    (b) False

3. What type of vaccine contains a live antigen?
   (a) Attenuated whole agent
   (b) Inactivated whole agent
   (c) Toxoids
   (d) Subunit

4. What type of vaccine uses fragments of a dead antigen?
   (a) Attenuated whole agent
   (b) Inactivated whole agent
   (c) Toxoids
   (d) Subunit

5. What type of vaccine uses a dead antigen killed by phenol or formalin?
   (a) Attenuated whole agent
   (b) Inactivated whole agent
   (c) Toxoids
   (d) Subunit

6. A conjugated vaccine is used for children who are under the age of 24 months.
   (a) True
   (b) False

7. What genetic engineering process is used to create a vaccine?
   (a) Toxoid
   (b) Recombinant
   (c) Active whole agent
   (d) Inactive whole agent

8. What test uses electrical current in a gel to increase the reaction time between an antigen and antibody?
   (a) Electrophoresis
   (b) Electrosynthesis
   (c) Gel separation
   (d) Gel division

9. A toxoid is an active toxin.
   (a) True
   (b) False

10. A booster is required for all immunization.
    (a) True
    (b) False

Antimicrobial Drugs

Probably the first thing you ask for when you're sick is a pill to make you feel better. Typically you don't care how the pill works as long as it does work.

Physicians have an arsenal of substances that combat disease. These substances are chemical agents called *drugs* and taking drugs to cure a disease is called *chemotherapy*. Throughout this book you learned about *pathogenic microorganisms* that are harmful when they invade a body. However, there are some microorganisms that actually prevent the harmful effect of pathogenic microorganisms.

A microorganism that destroys a pathogenic microorganism and used to cure a disease is referred to as an *antimicrobial drug*. You'll learn about antimicrobial drugs in this chapter.

Chemotherapeutic Agents: The Silver Bullet

When we hear the term chemotherapy, we tend to think of an ongoing treatment for cancer. While this is true, chemotherapy is any treatment that introduces a chemical substance into the body to destroy a pathogenic microorganism. The chemical substance is a *chemotherapeutic agent*.

The chemotherapeutic agent must do two things. First, it must only cause minimal damage (or no damage) to host tissues. A *host* is the term scientists use to refer to the patient who is receiving chemotherapy; *host tissues* are tissues of the patient's body. This simply means that the chemotherapeutic agent must not injure the patient or, if there is injury, that the injury is minimal and the patient's body will regenerate the destroyed tissues once chemotherapy is finished.

The second thing that the chemotherapeutic agent must do is to destroy the pathogenic microorganism that is causing the disease. The way in which a chemotherapeutic agent destroys a pathogenic microorganism is called the chemotherapeutic agent's *action*. The pathogenic microorganism that is attacked by the chemotherapeutic agent is called the chemotherapeutic agent's *target*.

There is generally one of two actions that a chemotherapeutic agent takes when combating a target. One is to kill the pathogenic microorganism outright, which is referred to as *bactericidal* action. The other is to inhibit the growth of the pathogenic microorganism, which is called *bacteriostatic* action. You'll learn more about bactericidal and bacteriostatic throughout this chapter.

A LOOK BACK

The idea of chemotherapy was the brainchild of Paul Ehrlich, a Germany scientist, who in the early twentieth century predicted that chemotherapeutic agents could be used to treat diseases that were caused by microorganisms. Ehrlich based his prediction on the result of a wayward experiment. He tried to stain only the bacteria in a tissue sample without staining the tissue.

This became a major hurdle in the discovery of chemotherapeutic agents. Finding a chemotherapeutic agent to kill a pathogenic microorganism wasn't difficult, but chemotherapeutic agents also harmed and sometimes killed the patient, too. The challenge was to discover a chemotherapeutic agent that cured the disease and not kill or severely injure the patient.

A breakthrough came in 1929 when Alexander Fleming was growing *Staphylococcus aureus*, a bacterium, in a Petri dish. A colony of a mold that contaminated the Petri dish surrounded the *Staphylococcus aureus* and prevented the bacterium from growing. The mold was *Penicillium notatum*. Fleming was able to isolate the part of the *Penicillium notatum* that stopped the growth of *Staphylococcus aureus*, which is referred to as the *active compound*. Fleming named this active compound *penicillin*.

The action of a compound to inhibit the growth of a microorganism is called antibiosis. From the word antibiosis comes the word antibiotic, which is any substance that a microorganism produces, in small amounts, that inhibits the growth of another microorganism. Therefore, penicillin is an antibiotic produced by the *Penicillium notatum* microorganism that inhibits the growth of the *Staphylococcus aureus* bacterium.

It took ten years since Fleming discovered penicillin before the first clinical trials were successful. These clinical trials proved to everyone that penicillin cured diseases caused by the *Staphylococcus aureus* bacterium. The next challenge was to mass-produce penicillin. This required a highly productive strain of *Penicillium*. The break-through came with the isolation of a highly productive strain of *Pennicillium* isolated from a cantaloupe fruit.

Many of the antibiotics in use today are produced from the *Streptomyces* species. *Streptomyces* are bacteria that live in the soil. Other antibiotics come from the genus *Bacillus* bacteria and the genera *Cephalosporium* and *Penicillium* which are molds.

Antimicrobial Activity: Who to Attack?

The way in which an antibiotic attacks a pathogenic microorganism is called the *antimicrobial activity*. You might say this is the way the antibiotic identifies the good guys from the bad guys. The good guys are eukaryotic cells and the bad guys are prokaryotic cells (bacteria). Human cells do not chemically resemble bacteria cells, this is the reason why the antibiotic can differentiate between the good guys and the bad guys.

Eukaryotic cells and prokaryotic cells are different in a number of ways, such as the presence or absence of cell walls and their chemical make up. There are also differences in their respective metabolisms and structures of their organelles, such as the ribosome. It is these differences that antibiotics target so that only the prokaryotic cells are destroyed.

Sometimes there can be a problem, especially when the disease-causing agents are other eukaryotic cells. While bacteria cells are dissimilar to human cells, the same cannot be said of other pathogenic microorganisms such as helminths, fungi, and protozoa. These microorganisms are comprised of eukaryotic cells that resemble human cells. Therefore antibacterial agents are not effective against helminths, fungi, and protozoa. Likewise, antibiotics are of no use against viruses because a virus invades a human cell and reprograms the human cells with the virus' genetic information to create new viruses. Since the virus is inside the human cell, the antibiotic is ineffective.

SPECTRUM OF ANTIMICROBIAL ACTIVITY: SOME BAD GUYS GET AWAY

The number of different types of pathogenic microorganisms that an antibiotic can destroy is called the *spectrum of antimicrobial activity*. These are referred to as broad-spectrum antibiotic or narrow-spectrum antibiotic.

A *broad-spectrum antibiotic* is an antibiotic that destroys many types of bacteria, such as both gram-positive and gram-negative bacteria. A *narrow-spectrum antibiotic* is an antibiotic that destroys a few types of bacteria, such as only gram-negative bacteria.

The deciding factor in the spectrum of antimicrobial activity is porins in the lipopolysaccharide outer layer of gram-negative bacteria. A *porin* is a water-filled channel that forms in the lipopolysaccharide outer layer, enabling substances on the outside of the cell to enter the cell.

In order for an antibacterial drug to destroy the bacteria the drug must enter the bacteria cell through the porin channel. However, to do so, the drug must be relatively small and hydrophilic. *Hydrophilic* means that the antibacterial drug has an affinity for water, which is contained in the porin channel. Some drugs are relatively large or are lipophilic. *Lipophilic* means that the antibiotic has an affinity for lipids and is attracted to the lipopolysaccharide outer layer of the cell (rather than the water in the porin channel).

THE BATTLE OF THE PATHOGENS: SOME BAD GUYS ARE GOOD GUYS TOO

Our bodies contain many microorganisms that are normal and beneficial. Others simply are unable to grow to the level where they become pathogenic because they compete with other microorganisms for nutrients required for growth.

This situation causes scientists concern when giving a broad-spectrum antibiotic to a patient when the pathogen is not known. A *pathogen* is a disease-causing organism. A broad-spectrum antibiotic is likely to destroy the pathogen, but it is also likely to destroy other microorganisms. This could cause an imbalance with competing microorganisms, resulting in a competitor being killed. This in turn enables the surviving microorganism to become an opportunistic pathogen. The increased growth of opportunistic pathogens is called *superinfection*. Microorganisms that develop resistance to the antibiotic also cause a *superinfection* by replacing the antibiotic-sensitive strain.

The Attack Plan

As previously discussed in this chapter, an antimicrobial drug uses one of two strategies to combat a pathogen. These are bactericidal or bacteriostatic. The bactericidal strategy is a direct hit, killing the pathogen and preventing it from spreading. Baceriostatic prevents the growth of microorganisms. In bacteriostasis, the host's immune system fights the pathogen through phagocytosis and the production of antibodies.

THE BACTERIOSTATIC STRATEGY

One of the first targets of attack of the bacteriostatic strategy is the cell wall of the pathogen. The objective is to weaken the cell wall, causing the cell to undergo lysis. The key to this attack is the structure of the cell wall itself. Bacteria cell walls are comprised of a network of marcromolecules called *peptidoglycan*. Certain antibiotics inhibit the making of peptidoglycan (synthesis), thus weakening the cell wall. Antibiotics that affect the synthesis of the cell wall of bacteria are bacitracin, vancomycin, penicillin, and cephalosporins.

Attacking Protein Synthesis

Another target of attack of the bacteriostatic strategy is the pathogen's capability to make protein. Protein is necessary for both eukaryotic and prokaryotic cells. If the antibiotic can inhibit protein synthesis, then the cell dies. The problem is for the antibiotic to identify only prokaryotic cells (bacteria) and not eukaryotic, which includes human cells.

The solution lies within the structure of ribosomes in eukaryotic and prokaryotoic cells. Ribosomes are the sites of protein synthesis.

Eucaryotic cells have 80S ribosomes and procaryotic cells have 70S. The numbers are Svedberg units and indicate the relative sedimentation rates during centrifugation. Prokaryotic ribosomes consist of a small subunit referred to as 30S and a large subunit called 50S. The 30S subunit contains one molecule of rRNA and the 50S subunit contains two molecules of rRNA.

Antibiotics use the differences in ribosomes to identify the prokaryotoic cell from the eukaryotic cell, thereby interfering with protein synthesis only for the prokaryotoic cells. Some antibiotics interfere with the 50S subunit while other antibiotics attack the 30S subunit. These antibiotics include chloramphenicol, erythromycin, streptomycin, and tetracycline.

Chloramphenicol interferes with the 50S subunit by preventing peptide bonds from forming. *Erythromycin* is a narrow-spectrum antibiotic that also interferes with the 50S subunit but only for gram-positive bacteria. *Tetracycline* interferes with the 30S subunit and prevents the tRNA from carrying amino acids and prevents amino acids from attaching to the polypeptide chain. *Streptomycin* interferes with the 30S subunit by changing its shape, causing an incorrect reading of the genetic code on the mRNA. Streptomycin is an example of what is called an *aminoglycoside* antibiotic. (An aminoglycoside is made up of amino carbohydrates and an aminocyclitol ring).

Attacking Plasma Membrane

Still another target of attack of antibiotics is the pathogen's plasma membrane. The plasma membrane is permeable, allowing substances in and out of the cell as part of normal cell metabolism.

Some antibiotics change the permeability of the plasma membrane, thereby disrupting the metabolism of the pathogen. One such antibiotic is polymyxin B, which attaches to the phospholipids of the plasma membrane, inhibiting permeability of the membrane.

Antifungal drugs also destroy fungus using a similar technique the cell membranes of fungi are made predominately of *erogosterol*. The plasma membrane of fungi have ergosterol. Antifungal drugs combine with sterols to inhibit the permeability of the membrane. Popular antifungal drugs include amphotericin B, ketoconazole, and miconazole.

Attacking Synthesis of Nucleic Acid

Nucleic acids are blueprints for the reproduction of every cell; these are DNA and RNA. A commonly used bacteriostatic strategy is to interfere with the making of nucleic acid by using rifampin, quinolones, and other similar antibiotics.

Although disrupting the formation of nucleic acid results in the destruction of the pathogen, scientists are careful when choosing the antibiotic for this purpose because the antibiotic might interfere with the host's DNA and RNA.

Attacking Metabolities

A *metabolite* is a substance (such as an enzyme) that is necessary for cell metabolism. The bacteriostatic strategy interferes with a metabolite and prevents the growth of the pathogenic organism.

For example, pathogens require the substrate para-aminobenzoic acid (PABA) in order to synthesize folic acid. Folic acid is a coenzyme that is involved in the synthesis of purine and pyrimidine nucleic acid bases and many amino acids.

The antimetabolite called *sulfanilamide*, which is a sulfa drug, resembles PABA. When sulfanilamide is introduced into the pathogenic microorganism, the enzyme used in making folic acid combines with sulfanilamide instead of PABA. This then disrupts the formation of folic acid and eventually prevents the pathogenic microorganism from synthesizing purine and pyrimidine.

Exploring Antimicrobial Drugs

Antimicrobial drugs are classified by their antimicrobial activity. These classifications are cell wall inhibitors, protein inhibitors, plasma membrane inhibitors, nucleic acid inhibitors, antimetabolites, and antifungal drugs, antiviral drugs, and enzyme inhibitors, antiprotozoan drugs, and antihelminthic drugs.

Let's take a close look at each of these classifications.

CELL WALL INHIBITORS

A cell wall inhibitor is an antimicrobal drug that inhibits the growth or functionality of the cell wall of a pathogen. The more common cell wall inhibitors follow:

Penicillin

Penicillin is a group of antibiotics that have a *β-lactam ring* the core structure of penicillins. This core structure is referred to as a "nucleus." Each member of the group has a different side chain that is attached to the *β*-lactam ring. Penicillin can be natural or semisynthetic.

- *Natural penicillin.* Natural penicillin is extracted from the *Penicillium* mold. The major disadvantage of natural penicillin, except for Penicillin V, is that it is negatively affected by stomach acid, meaning that the most effective way to administer natural penicillin is an intramuscular injection. Another problem is that penicillin itself can be attacked by the enzyme penicillinase. Penicillinases, also known as *β*-lactamases, are enzymes produced by many bacteria that attach to the *β*-lactam ring of the penicillin, rendering penicillin ineffective. The more common natural penicillins are:

- *Penicillin G* is the prototype compound for natural penicillin as is used in defense of staphylococci, streptococci, and several spirochetes. Natural penicillin has a narrow spectrum of activity. Penicillin G lasts up to six hours when injected intramuscularly.
- *Procaine penicillin* is a combination of the drugs procaine and penicillin G and lasts up to 24 hours, although the concentration of procaine penicillin diminishes after four hours.
- *Benzathine penicillin* is a combination of benzathine and penicillin G and can last for up to four months.
- *Penicillin V* is not inhibited by acid in the stomach and can be taken orally.
- *Semisynthetic penicillin*. Semisynthetic penicillins are chemically altered natural penicillins that are designed to overcome disadvantages of natural penicillins. There are two ways in which scientists modify natural penicillins. They stop the natural synthesis of a *Penicillium* molecule and use the penicillin nucleus. The other way is to modify the side chains from the natural penicillin and insert a new side chain that overcomes the disadvantage of the original side chain. This is the technique used to combat penicillinase. Common types of semisynthetic penicillin are:
  - *Methicillin* was the first semisynthetic penicillin developed to resist penicillinase.
  - *Oxacillin* is a newer semisynthetic penicillin that resists penicillinase and has replaced methicillin.
  - *Ampicillin* is a semisynthetic penicillin that overcomes the narrow spectrum of activity of natural penicillin by attacking gram-negative and gram-positive bacteria. Ampicillin is not resistant to penicillinases.
  - *Amoxicillin* is similar to ampicillin.
  - *Carbenicillin* is a member of the carboxypenicillin group and has a broad spectrum of activity against gram-negative bacteria and is used to fight *Pseudomonas aeruginosa*.
  - *Ticarcillin* is similar to carbenicillin.
  - *Mezlocillin* is a member of the ureidopenicillins group of penicillins and is a modification of ampicillin, giving it a broader spectrum of activity against bacteria.
  - *Azlocillin* is similar to mezlocillin.
  - *Augmentin* (which is the trade name) is a combination of amoxicillin and potassium clavulanate, which is produced by streptomycete. Amoxicillin attacks gram-negative and gram-positive bacteria and potassium clavulanate resists penicillinase.

Monobactams

Monobactams are synthetic antibiotics that have a single lactam ring and resist penicilinase. Monobactams destroy only areobic gram-negative bacteria such as *E. coli.*

Cephalosporins

Cephalosporins are resistant to penicillinases and destroy gram-negative bacteria. The disadvantage of cephalosporins is they can be made ineffective by β-lactamases.

Carbapenems

Carbapenems have a broad spectrum of activity that prevents the breakdown of the antibiotic by the kidneys. These antibiotics inhibit the synthesis of cell walls. An example is Primax (trade name). Primax has proved effective against 98 percent of all organisms that were isolated from patients in hospitals.

Bacitracin

Bacitracin is used against staphylococci and streptococci and other gram-positive bacteria and is used as a topical antibiotic for superficial infections.

Vancomycin

Vancomycin has a very narrow spectrum of activity and is the most effective antibiotic against staphylococci that produce penicillinase. Vancomycin is used in the treatment of endocarditis. However, vancomycin can be toxic to humans.

Isoniazid

Isoniazid (INH) inhibits mycolic acid, which is needed for synthesizing the cell wall of *Mycobacteria tuberculosis*. Isoniazid is only effective fighting mycobacteria tuberculosis.

Ethambutol

Ethambutol is similar to isoniazid and is used as a secondary treatment to avoid *Mycobacteria tuberculosis* becoming resistant to isoniazid.

PROTEIN INHIBITORS

A protein inhibitor drug interferes with the pathogen's capability to synthesize protein. The commonly used protein inhibitor drugs follow.

Aminoglycosides

Aminoglycosides are a group of antibiotics that were the first to attack gram-negative bacteria. Their disadvantage is that they can cause permanent damage to the auditory nerve and cause kidney damage. Commonly used members of this group of antibiotics are:

- *Streptomycin* is used as a secondary treatment to tuberculosis. It was discovered in 1944 and today is used sparingly because of its toxic effect on humans and the fact that bacteria quickly become resistant to it.
- *Neomycin* is a topical antibiotic used for superficial infections.
- *Gentamicin* is an antibiotic used to combat *Pseudomonas aeruginosa* infections.

Tetracyclines

Tetracycline is a broad-spectrum antibiotic that attacks both gram-positive and gram-negative bacteria. Tetracyclines are able to combat pathogens that invade cells because they can enter body tissues. They are used against urinary tract infections, rickettsial infection, and chlamydial infections. Tetracyclines are also used as the secondary treatment for gonorrhea and syphilis. The disadvantages of tetracyclines are discoloration of teeth in children and liver damage in pregnant women. Common members of the tetracycline group of antibiotics are:

- *Oxytetracycline* is a commonly used form of tetracycline that is also known as Terramycin.
- *Chlortetracyline* is another commonly used form of tetracycline that is also known as Aureomycin.
- *Doxycycline* is a semisynthetic form of tetracycline that has a longer-lasting effect than the natural form of tetracycline.
- *Minocyline* is a semisynthetic form of tetracycline.

Chloramphenicol

Chloramphenicol is a broad-spectrum antibiotic that is small enough to effect areas of the body that are too small for other antibiotics to enter. Chloramphenicol is an antibiotic of last resort because it inhibits formation of blood cells, causing aplastic anemia.

Macrolides

Macrolides are a group of antibiotics that have a similar effect as that of penicillin G. They affect the organism by inhibiting its ability to synthesize protein and are a secondary drug when penicillin G is unavailable or the organism has become resistant to penicillin G. Macrolide antibiotics can be administered orally, making them the choice for treating children who have streptococcal and staphylococcal infections. The most commonly used macrolide is *Erythromycin,* which is used to treat legionellosis, mycoplasmal pneumonia, and streptococcal and staphylococcal infections.

PLASMA MEMBRANE INHIBITORS

Plasma membrane inhibitors are antibiotics that interfere with the functionality of the plasma membrane of a pathogen. The commonly used plasma membrane inhibitor antibiotic is *Polymyxin B*, which combats gram-negative bacteria such as *Pseudomonas*. Today, when combined with neomycin, it is used in nonprescription topical antibiotics for superficial infections.

NUCLEIC ACID INHIBITORS

Nucleic acid inhibitors interfere with the formation of nucleic acids. The commonly used nucleic acid inhibitor antibiotics follow.

Rifamycins

Rifamycins comprise a group of antibiotics that inhibit mRNA synthesis. They are used to treat tuberculosis and leprosy. Rifamycins can reach the cerebrospinal fluid and enter tissues and abscesses. The disadvantage of rifamycins is that they cause urine, feces, saliva, sweat and tears to appear an orange-red

color. A commonly used rifamycin is *rifampin*, which is used to attack myco-bacteria that cause tuberculosis and leprosy.

Quinolones

Quinolones interfere with DNA gyrase (an enzyme), which is involved in DNA replication. Quinolones are used only for infections of the urinary tract.

Fluoroquinolones

Fluoroquinolones comprise a group of antibiotics that have a broad-spectrum of activity and target urinary tract infections. Fluoroquinolones are able to enter cells to attack pathogens. The disadvantage of fluoroquinolones is that they may interfere with cartilage development, making them unsafe for pregnant women and children. Two examples that are mostly used are *norfloxacin* and *ciprofloxacin.*

ANTIMETABOLITES

Antimetabolites are a group of drugs that interfere with metabolite, which is a substance (such as an enzyme) that is necessary for cell metabolism. Common antimetabolites follow.

Sulfonamides

Sulfonamides or sulfa drugs are used to treat urinary tract infections and control infections in burn patients. These were among the first synthetic antimicrobial drugs used. Commonly used sulfonamides are:

- *Silver sulfadiazine* is used to control infections common in burn patients.
- TMP-SMZ is a combination of trimethoprim and sulfamethoxazole and is used to combat intestinal tract and urinary tract gram-negative pathogens.

ANTIFUNGAL DRUGS

Antifungal drugs inhibit the growth of fungi. The commonly used antifungal drugs follow:

Polyenes

Polyenes are a group of antifungal antibiotics that combine with the ergosterol in the plasma membrane of fungi. This causes the plasma membrane to become extremely permeable, ultimately resulting in the death of the fungi. A commonly used polyene is *amphotericin B,* which is used to treat histoplasmosis, coccidioidomycosis, and blastomycosis. However, amphotericin B is toxic to kidneys. Enclosing amphotericin B in liposomes when administering the drug reduces its toxicity.

Imidazoles

Imidazoles interfere with fungal ergosterol synthesis. Commonly used imidazoles are:

- *Clotrimazole* is used to treat cutaneous mycoses. Two example are athlete's foot and vaginal yeast infections.
- *Miconazole* is similar to clotrimazole. Both drugs are used topically and sold without a prescription.

Triazoles

Triazoles are similar to imidazoles, but are less toxic than other antifungals.

Griseofulvin

Griseofulvin is an antibiotic used to treat dermatophytic fungal infections of the hair and nails. This includes tinea capitis. Griseofulvin is taken orally and enters the keratin of skin, hair, and nails. Its purpose is to interfere with fungal mitosis, inhibiting reproduction.

Tolnaftate

Tolnaftate is a topical treatment for athlete's foot. Its mechanism for action is unknown.

ANTIVIRAL DRUGS

Antiviral drugs interfere with the replication of viruses. Commonly used antiviral drugs follow.

Amantadine

Amantadine prevents a virus from entering the cell or from uncoating once the virus enters the cell. It is used (though it has limited usefulness) to prevent influenza A, but has no practical effect once the virus infects the cell.

Nucleoside analogs

Nucleoside analog antiviral drugs affect the synthesis of viral DNA or RNA. Commonly used nucleoside analogs are:

- *Acyclovir* is used to combat viruses that cause herpes.
- *Ribavirin* is used to treat rotavirus-caused pneumonia in infants.
- *Ganciclovir* is used fight cytomegalovirus infections that are common in transplant patients and patients who have AIDS.
- *Trifluridine* is used to treat the eye infection caused by acyclovir-resistant herpes keratitis.
- *Zidovudine* (AZT) is used in HIV infection. This drug blocks the synthesis of DNA from RNA by the enzyme reverse transcriptase.

ANTIPROTOZOAN DRUGS

Antiprotozoan drugs are used to combat parasitic infection, such as the disease malaria. Commonly used antiprotozoan drugs follow.

Chloroquine

Chloroquine is a replacement for quinine, which was the traditional treatment for malaria.

Mefloquine

Mefloquine is a secondary treatment for malaria when there is a resistance to *chloroquine*.

Quinacrine

Quinacrine is used to treat giardiasis.

Diiodohydroxyquin

Diiodohydroxyquin is used to treat intestinal amoebic diseases. However, *diiodohydroxyquin* can cause damage to the optic nerve if the dosage is not carefully controlled.

Metronidazole

Metronidazole combats parasitic protozoa and obligate anaerobic bacteria. It is used to treat vaginitis caused by *Trichomonas vaginalis,* giardiasis, and amoebic dysentery.

Nifurtimox

Nifurtimox is used to combat Chagas' disease. However, nifurtimox can cause side effects, such as nausea and convulsions.

ANTIHELMINTHIC DRUGS

Antihelminthic drugs are used to combat helminths, which are parasitic flatworms. Commonly used antihelminthic drugs follow.

Niclosamide

Niclosamide is used to treat tapeworm infections. This drug inhibits ATP production in aerobic conditions.

Praziquantel

Praziquantel is also used to destroy tapeworm infections and is used to treat schistosomiasis and other fluke-caused diseases. This drug alters the permeability of the organism's plasma membrane.

Mebendazole

Mebendazole is used to combat intestinal helminthic infections, such as ascariasis, whipworms, and pinworms.

Chemotherapy Tests

A *chemotherapy test* is a scientific test that determines which antibiotic combats a particular pathogen. These tests are used to determine which antibiotic to prescribe to treat a specific disease. The antibiotic is called a *chemotherapeutic agent*.

There are two popular chemotherapy tests: diffusion methods and the broth dilution method.

DIFFUSION METHODS

Diffusion methods are tests that place a paper disk or plastic strip that is coated with a chemotherapeutic agent in touch with a pathogen to determine if the chemotherapeutic agent inhibits the pathogen. There are two commonly used diffusion methods. These are the disk-diffusion method and the minimal inhibitory concentration method.

The *disk-diffusion method*, also known as the *Kirby-Bauer* test, uses a filter paper disk that is coated with the chemotherapeutic agent being tested. The paper disk is placed in a Petri plate that contains an agar medium that has been contaminated with the pathogen. The Petri plate is then incubated during which the chemotherapeutic agent diffuses onto the agar. A zone of inhibition can be seen in the agar medium. A zone of inhibition is the area where the pathogen does not grow because the chemotherapeutic agent inhibits it. Scientists measure the diameter of the zone of inhibition and compare the diameter to a standard table for the chemotherapeutic agent. The table indicates whether the pathogen is sensitive, intermediate, or resistant to the chemotherapeutic agent.

The *minimal inhibitory concentration* (MIC) test is an advanced diffusion method that determines the lowest concentration of the chemotherapeutic agent that inhibits the visible growth of bacteria. This test uses a plastic coated strip that contains a gradient concentration of the chemotherapeutic agent. The strip is then exposed to the pathogen, after which the strip is compared to a printed scale to determine the minimal inhibitory concentration of the chemotherapeutic agent.

BROTH DILUTION METHOD

The *broth dilution method* is used to determine the minimal inhibitory concentration and the minimal bactericidal concentration (MBC) of the chemotherapeutic agent. The minimal bactericidal concentration is the lowest concentration of a chemotherapeutic agent needed to kill a pathogen.

The broth dilution test requires that a broth containing the drug be placed in wells of a plastic tray in a sequence of decreasing concentrations of the drug. Each well is inoculated with the bacteria. After an incubation period, each well is examined to determine the effectiveness of the concentration of the chemotherapeutic agent. The well that shows no pathogen growth with the lowest concentration of the chemotherapeutic agent signifies the MIC and MBC that should be used to treat the disease caused by the pathogen.

Clinical laboratories used an automated broth dilution test where a computer scans wells and reports results.

Quiz

1. Antibiosis is
   (a) an inactive compound
   (b) an antibiotic
   (c) narrow-spectrum
   (d) broad-spectrum

2. Prokaryotic cells are human cells that are targeted by bacteria.
   (a) True
   (b) False

3. The way in which an antibiotic attacks a pathogenic microorganism is called
   (a) lipophilic
   (b) hydrophilic
   (c) lipopolysaccharide
   (d) antimicrobial activity

4. A superinfection is caused by
   (a) a pathogen that is resistant to the antibiotic
   (b) an infection in deep tissues of the body
   (c) an infection that covers a large area of the body
   (d) an infection that infects more than two organs

5. What destroys many types of bacteria?
   (a) Narrow-spectrum antibiotics
   (b) Broad-spectrum antibiotics
   (c) Porins
   (d) Hydrophilic

6. The bacteriostatic strategy is a direct hit, killing the pathogen.
   (a) True
   (b) False

7. What is the spectrum of antimicrobial activity?
   (a) Broad-spectrum antibiotic
   (b) Narrow-spectrum antibiotic
   (c) The minimum number of different types of pathogenic microorganisms that an antibiotic can destroy
   (d) The number of different types of pathogenic microorganisms that an antibiotic can destroy

8. Sulfanilamide is a
   (a) Cell wall inhibitor
   (b) Protein inhibitor
   (c) Antimetabolite
   (d) DNA inhibitor

9. The bacteriostasis strategy causes the host's immune system to fight the pathogen through phagocytosis.
   (a) True
   (b) False

10. Penicillin is a cell wall inhibitor.
    (a) True
    (b) False

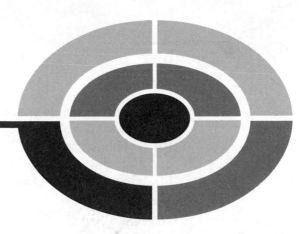

Final Exam

1. What microorganisms can be harmed by oxygen?
   (a) Obligate aerobes
   (b) Facultative anaerobes
   (c) Obligate anaerobes
   (d) Free radicals

2. An inoculating loop is used to insert a pathogen into the patient's mouth.
   (a) True
   (b) False

3. Which microorganisms can use oxygen when present and continue to grow without it?
   (a) Obligate aerobes
   (b) Facultative anaerobes
   (c) Obligate anaerobes
   (d) Free radicals

4. Facultative anaerobes require oxygen.
   (a) True
   (b) False

5. A medium containing no living organisms is called a
   (a) sterile culture medium
   (b) an obligate medium
   (c) a chemically defined medium
   (d) a moisture-filled medium

6. A chemically defined medium must contain growth factors.
   (a) True
   (b) False

7. What is agar?
   (a) A nutrient
   (b) A vitamin
   (c) A solidifying agent
   (d) A broth

8. What culture provides growth of a certain microorganism but not others?
   (a) Inherited culture
   (b) Inhibited culture
   (c) An infectious culture
   (d) An enrichment culture

9. How can preserved bacteria cultures be revived?
   (a) Oxygenation
   (b) Hydration
   (c) Hydration and liquid nutrient medium
   (d) Oxygenation and liquid nutrient medium

10. A microorganism is a small organism that intakes and breaks down food for energy and nutrients, excretes unused food as waste, and is capable of reproduction.
    (a) True
    (b) False

11. A pathogenic microorganism is a microorganism that causes disease in a host.
    (a) True
    (b) False

12. A prokaryotic organism is a one-celled organism that does not have a distinct nucleus.
    (a) True
    (b) False

13. Why is a fungus called a eukaryotic microorganism?
    (a) Because it has a nucleus, nuclear envelope, cytoplasm, and organelles.
    (b) Because it has a nucleus and no nuclear envelope.
    (c) Because it has a nucleus, nuclear envelope, and cytoplasm, but no organelles.
    (d) Because it has no nucleus, no nuclear envelope, no cytoplasm, and no organelles.

14. Phagocytosis is the process by which a cell engulfs and digests solid materials through the use of pseudopods or "false feet."
    (a) True
    (b) False

15. Germ Theory states that a disease-causing microorganism should be destroyed.
    (a) True
    (b) False

16. Adenisine triphosphate is the preferred energy-storage molecule.
    (a) True
    (b) False

17. What reaction requires energy as small molecules are combined to form large molecules?
    (a) Anabolic reaction
    (b) Catabolic reaction
    (c) Anabolism
    (d) Hydrolases reaction

18. What reaction releases energy as large molecules are broken down (metabolized) into small molecules?
    (a) Anabolic reaction
    (b) Catabolic reaction
    (c) Anabolism
    (d) Hydrolases reaction

19. Activation energy is the energy required to begin a chemical reaction.
    (a) True
    (b) False

20. What process is used to break down glucose?
    (a) Anabolic reaction
    (b) Catabolic reaction
    (c) Anabolism
    (d) Hydrolases reaction

21. What is it called when a molecule donates electrons?
    (a) Oxidation
    (b) Reduction
    (c) Oxidation reduction
    (d) Lactation

22. What is caused when there is a chemical denaturing of enzymes?
    (a) Protein creation
    (b) Lipid creation
    (c) Very high or very low pH
    (d) Saturation point is reached

23. Pyruvic acid blocks active sites from bonding to a substrate.
    (a) True
    (b) False

24. The spontaneous decaying and giving off of particles of radium is called
    (a) catalyst
    (b) chemically stable
    (c) half-life
    (d) radioactivity

25. A differential element is the name given to a chemical element whose atoms have a different number of neutrons.
    (a) True
    (b) False

26. The valence shell is used in bonding together two
    (a) stable atoms
    (b) inorganic stable atoms

(c) organic stable atoms

(d) unstable atoms

27. A reaction that breaks the bond between atoms of a molecule or compound is called a decomposition reaction.
    (a) True
    (b) False

28. What kind of reaction performs synthesis and decomposition?
    (a) Endergonic reaction
    (b) Fusion reaction
    (c) Fission reaction
    (d) Exchange reaction

29. The reaction rate is increased by
    (a) temperature
    (b) pressure
    (c) orientation
    (d) all of the above

30. The enzyme temporarily bonds to the substrate and then positions the substrate to increase the likelihood that the substrate will collide with another atom, ion, or molecule and thereby bring about a chemical reaction.
    (a) True
    (b) False

31. Describe a gram-positive cell wall.
    (a) A gram-positive cell wall has many layers that repels the crystal of violet dye when the cell is stained.
    (b) A gram-positive cell wall has many layers of peptidoglygan that retain the crystal of violet dye when the cell is stained.
    (c) A gram-positive cell wall has one layer that retains the crystal of violet dye when the cell is stained.
    (d) A gram-positive cell wall has one layer of peptidoglygan that repels the crystal of violet dye when the cell is stained.

32. The cytoplasmic membrane is a membrane that provides a selective barrier between the nucleus and the cell's internal structures.
    (a) True
    (b) False

33. What regulates the movement of molecules through the cytoplasmic membrane?
    (a) Channel proteins
    (b) Transmembrane proteins
    (c) Peripheral proteins
    (d) Fluid proteins

34. What forms channels in the cytoplasmic membrane and permits the flow of molecules through the cytoplasmic membrane?
    (a) Integral proteins
    (b) Transmembrane proteins
    (c) Peripheral proteins
    (d) Lipoproteins

35. What is it called when a substance moves from a higher-concentration region to a lower-concentration region?
    (a) Osmosis
    (b) Cytoplasmic diffusion
    (c) Facilitated diffusion
    (d) Simple diffusion

36. What is it called when water moves from a higher region of concentration to a region of lower concentration?
    (a) Osmosis
    (b) Cytoplasmic diffusion
    (c) Facilitated diffusion
    (d) Simple diffusion

37. Passive transport is the movement of a substance across the cytoplasmic membrane against the gradient by using energy provided by the cell.
    (a) True
    (b) False

38. What are the two types of exocytosis?
    (a) Facilitated diffusion and passive transport
    (b) Phagocytosis and pinocytosis
    (c) Hydrolysis and cytosol
    (d) None of the above

39. What is a nanometer?
    (a) 1/1,000,000,000 of a meter
    (b) 1/100,000 of a meter
    (c) 1/1,000,000 of a meter
    (d) 1,000,000,000 meters

40. Immersing a specimen in oil maintains good resolution at magnifications greater than 100×.
    (a) True
    (b) False

41. A photograph taken by an electron microscope is called
    (a) a TEM
    (b) a micrograph
    (c) a SEM
    (d) a palladium

42. You stain a smear using
    (a) crystal violet
    (b) caboxymethyl cellulose
    (c) cellulose
    (d) a fixation process

43. Why is a specimen stained?
    (a) A stain is used to label a specimen.
    (b) A stain is used to determine the size of a specimen.
    (c) A stain adheres to the specimen, causing more light to be reflected by the specimen into the microscope.
    (d) A stain is used to determine the density of a specimen.

44. You use a wet mount to observe an inorganic specimen under a microscope.
    (a) True
    (b) False

45. The naming convention used to group organisms is called
    (a) identification
    (b) nomenclature
    (c) systemics
    (d) classification

46. Fungi have a cell wall composed of peptidoglycan and muramic acid.
    (a) True
    (b) False

47. Identification is the process of observing and classifying organisms into a standard group.
    (a) True
    (b) False

48. A genus consists of one or more lower ranks called species.
    (a) True
    (b) False

49. Which group acquires nutrients through absorption?
    (a) animals
    (b) plants
    (c) fungi
    (d) humans

50. Taxonomy is the classification of organisms based on a scientifically proven natural relationship.
    (a) True
    (b) False

51. Which group acquires nutrients through photosynthesis?
    (a) animals
    (b) fungi
    (c) humans
    (d) none of the above

52. The arrangement of organisms into groups based on similar characteristics is called
    (a) nomenclature
    (a) identification
    (c) classification
    (d) systemics

53. The Southern blot technique is used for detecting specific restriction fragments.
    (a) True
    (b) False

54. What enables scientists to take nucleotide fragments from other DNA and reassemble fragments into a new nucleotide sequence?
    (a) enzyme DNA technology
    (b) enzyme technology
    (c) recombinant DNA technology
    (d) recombinant enzyme technology

55. The result of two incisions made in a double-helical segment is called
    (a) overhang
    (b) recognition sequence
    (c) restriction enzymes
    (d) restriction fragment

56. Restriction enzymes are used to cut double stranded DNA along the exterior of the strand.
    (a) Truc
    (b) False

57. What is another name for a restriction endonuclease?
    (a) plasmid
    (b) vector
    (c) agarose gel
    (d) restriction enzyme

58. The nucleotide genome consists of the seven nucleic acids that encode genetic information on RNA.
    (a) True
    (b) False

59. The end of the cut of a double-helical segment is called _____ in certain enzymes.
    (a) restriction enzymes
    (b) overhang
    (c) recognition sequence
    (d) restriction fragment

60. The four nucleotides are adenine (A), cytosine (C), guanine (G), and thymine (T).
    (a) True
    (b) False

61. Which of the following is not a type of RNA?
    (a) rRNA
    (b) uRNA
    (c) tRNA
    (d) mRNA

62. An operon consists of structural DNA, regulatory RNA, and control cells.
    (a) True
    (b) False

63. What process begins protein synthesis?
    (a) transcription
    (b) translation
    (c) polypeptide
    (d) genotyping

64. The point at which the double-stranded DNA molecule unwinds is called polymerase.
    (a) True
    (b) False

65. What is the promotor site?
    (a) A site where RNA polymerase binds to DNA.
    (b) A site where RNA polymerase binds to protein.
    (c) A site where RNA polymerase binds to free nucleotides.
    (d) A site where RNA polymerase binds to ATP.

66. The unexpressed properties of heredity are called the
    (a) genotype
    (b) exons
    (c) introns
    (d) phenotype

67. An RNA polymerase is an enzyme that is used in the synthesis of RNA.
    (a) True
    (b) False

68. The three nucleotide units withing the mRNA strand are called exons.
    (a) True
    (b) False

69. An incubatory carrier is an individual who transmits the disease before becoming symptomatic.
    (a) True
    (b) False

70. The number of people infected by a disease at any point in time is called the
    (a) incidence
    (b) morbidity rate
    (c) prevalence
    (d) none of the above

71. Morbidity rate is the number of people is a given population who are ill.
    (a) True
    (b) False

72. When there are small isolated occurrences of a disease reported it is called a
    (a) pandemic
    (b) source epidemic
    (c) sporadic disease
    (d) propagated epidemic

73. A disease is called an epidemic when there is a worldwide distribution of the disease.
    (a) True
    (b) False

74. When large numbers of a population are infected suddenly from a common source it is called a
    (a) pandemic
    (b) sporadic disease
    (c) propagated epidemic
    (d) none of the above

75. The carrier period is the period of time when a disease is contagious and communicable before showing signs and symptoms and during recovery.
    (a) True
    (b) False

76. The sum of old and new cases of a disease is called
    (a) incidence
    (b) prevalence rate
    (c) morbidity rate
    (d) prevalence

77. Convalescent carriers are individuals who develop chronic infections and transmit them for long periods of time.
    (a) True
    (b) False

78. A lytic virus injects nucleic acid into a host cell.
    (a) True
    (b) False

79. A bacteriophage is
    (a) a virus that can be killed by antibiotics
    (b) a virus that acts like a bacterium
    (c) a bacterium that acts like a virus
    (d) none of the above

80. A small infectious particle that contains a protein is called a
    (a) viroid
    (b) capsid
    (c) viron
    (d) none of the above

81. What is a capsid?
    (a) A capsid is the membrane bilayer of a virus.
    (b) A capsid is the protein coat that encapsulates a virus.
    (c) A capsid is another name for a bacteriophage.
    (d) A capsid is the envelope around a virus.

82. A free virus particle is
    (a) another name for a virus
    (b) a synthesized virus
    (c) a viron
    (d) a mature virus

83. Pieces of the host cell's membrane make
    (a) the capsid
    (b) the envelope of a virus
    (c) the nuclei of the virus
    (d) the nuclei of the host cell

84. The central nucleic acid contains the genetic information of a virus.
    (a) True
    (b) False

85. Oncogenic viruses
    (a) cause herpes
    (b) cause the common cold
    (c) cause tumors
    (d) none of the above

86. A nakcd virus is a virus without a capsid.
    (a) True
    (b) False

87. A state of lysogeny occurs when the envelope of a virus interacts with the capsid.
    (a) True
    (b) False

88. What are protists?
    (a) hemoflagellates
    (b) nematodes
    (c) proglottids
    (d) scolex

89. Mushrooms are
    (a) protozoa
    (b) fungi
    (c) helminths
    (d) algae

90. Which is not a layer of the body of helminths?
    (a) mesoderm
    (b) ectoderm
    (c) exoderm
    (d) endoderm

91. Stipes support
    (a) blades
    (b) thallus
    (c) algae
    (d) photoautotrophs

92. What is the method used by multicellular algae to reproduce?
    (a) cytokinesis
    (b) scolex
    (c) nematodes
    (d) none of the above

93. Fungi are different from plants because plants absorb nutrients from organic matter.
    (a) True
    (b) False

94. Some protists use cilia to move food into a mouth-like opening called
    (a) pseudopods
    (b) mesoderm
    (c) vacuole
    (d) none of the above

95. What are flatworms?
    (a) platyhelminths
    (b) nematodes
    (c) ringhelminths
    (d) hookhelminths

96. A helminth is an alga.
    (a) True
    (b) False

97. A pneumatocyst is
    (a) a gas-filled bladder that enables algae to float
    (b) a way for protozoa to absorb nutrients
    (c) something that anchors fungi to rocks
    (d) something that enables helminths to reproduce

98. A serum that contains most antibodies is called gamma globulin.
    (a) True
    (b) False

99. What kind of immunity uses specialized lymphocyte cells called T cells?
    (a) antibody-mediated immunity
    (b) epitopes
    (c) cell-mediated immunity
    (d) none of the above

100. Immunity in which antibodies are developed outside the organism in an immune serum is called active naturally acquired immunity.
    (a) True
    (b) False

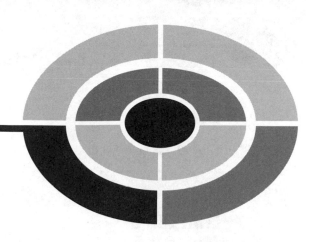

Answers to Quiz and Exam Questions

CHAPTER 1

1. a   2. d   3. d   4. a   5. a
6. c   7. c   8. b   9. a   10. b

CHAPTER 2

1. d   2. c   3. d   4. c   5. b
6. b   7. a   8. a   9. a   10. c

CHAPTER 3

1. a   2. a   3. c   4. d   5. c
6. a   7. a   8. a   9. c   10. b

CHAPTER 4

1. b   2. d   3. a   4. b   5. c
6. c   7. b   8. c   9. c   10. b

CHAPTER 5

1. a   2. b   3. a   4. b   5. b
6. b   7. c   8. a   9. c   10. c.

CHAPTER 6

1. b   2. a   3. c   4. b   5. b
6. a   7. c   8. d   9. a   10. c

CHAPTER 7

1. a   2. b   3. d   4. b   5. a
6. a   7. b   8. d   9. a   10. b

CHAPTER 8

1. a   2. c   3. b   4. d   5. a
6. c   7. a   8. c   9. a   10. a

CHAPTER 9

1. a   2. c   3. a   4. b   5. d
6. a   7. c   8. b   9. a   10. a

CHAPTER 10

1. a   2. c   3. d   4. a   5. b
6. a   7. c   8. d   9. a   10. b

CHAPTER 11

1. d   2. b   3. a   4. b   5. c
6. a   7. a   8. c   9. c   10. b

CHAPTER 12

1. b   2. a   3. d   4. d   5. a
6. b   7. b   8. a   9. d   10. b

CHAPTER 13

1. b   2. d   3. c   4. a   5. d
6. c   7. b   8. b   9. a   10. b

CHAPTER 14

1. b   2. a   3. c   4. b   5. d
6. b   7. a   8. b   9. a   10. a

CHAPTER 15

1. c   2. b   3. a   4. d   5. b
6. a   7. b   8. a   9. a   10. b

CHAPTER 16

1. b   2. b   3. d   4. a   5. b
6. a   7. d   8. c   9. a   10. a

FINAL EXAM

1. c   2. b   3. b   4. b   5. a
6. a   7. c   8. d   9. c   10. a

11. a	12. a	13. a	14. a	15. b
16. a	17. a	18. b	19. a	20. b
21. a	22. c	23. b	24. d	25. b
26. d	27. a	28. d	29. d	30. a
31. b	32. a	33. b	34. a	35. d
36. a	37. b	38. d	39. a	40. a
41. b	42. d	43. c	44. b	45. b
46. b	47. a	48. a	49. c	50. b
51. d	52. c	53. a	54. c	55. d
56. a	57. d	58. b	59. b	60. a
61. b	62. b	63. b	64. b	65. a
66. a	67. a	68. b	69. a	70. c
71. a	72. c	73. b	74. d	75. b
76. b	77. b	78. a	79. d	80. d
81. b	82. c	83. b	84. a	85. c
86. b	87. b	88. a	89. b	90. c
91. c	92. a	93. b	94. d	95. a
96. b	97. a	98. a	99. c	100. b

INDEX

INDEX

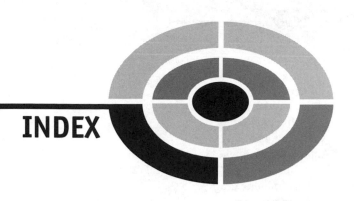

ABOUT THE AUTHORS

Tom Betsy, D.C., teaches microbiology, anatomy, physiology, and other courses in the nursing and allied health programs at Bergen Community College in northern New Jersey. Dr. Betsy is widely admired by his students for his ability to simplify complex subjects. He lives in Hillsdale, New Jersey.

Jim Keogh is a writer and teacher of computing at Columbia University. He is the author of more than 60 books. This is the seventh book in the *Demystified* series that he has authored. The others are *Pharmacology Demystified, JavaScript Demystified, ASP.NET Demystified, Data Structures Demystified, OOP Demystified, and Java Demystified.* He lives in Ridgefield Park, New Jersey.